# Interactions in Multiagent Systems: Fairness, Social Optimality and Individual Rationality

Jianye Hao • Ho-fung Leung

# Interactions in Multiagent Systems: Fairness, Social Optimality and Individual Rationality

Jianye Hao
School of Computer Software
Tianjin University
Tianjin, China

Ho-fung Leung
Department of Computer Sci. and Eng.
The Chinese University of Hong Kong
Hong Kong, China

ISBN 978-3-662-49468-4
ISBN 978-3-662-49470-7 (eBook)
HEP ISBN 978-7-04-044111-6
DOI 10.1007/978-3-662-49470-7

Library of Congress Control Number: 2016933452



Printed on acid-free paper

This Springer imprint is published by Springer Nature
The registered company is Springer-Verlag GmbH Berlin Heidelberg

# Preface

It has long been a fundamental assumption in the research of autonomous agents and multiagent systems that intelligent agents should be rational—well, at least as rational as possible even if they are handicapped, like lacking access to all information that is needed for rational decision-making. While this assumption is only natural and right, rationality is actually only a better word to use if we try to avoid the use of the word selfishness to describe the deed of concerning only with one's own interests and being regardless of that of others.

When we first decided to work on the issues of fairness and social optimality in multiagent systems, with due respect to individual rationality of agents, we knew that we were facing a nontrivial challenge. The first difficulty that we encountered was the definition of fairness. We did spend some time to investigate this issue, delving into the definitions given by various authoritative dictionaries, hoping to obtain some mathematical formalism that captured the essence of these definitions. Unfortunately, the definitions in most of these dictionaries were not too helpful to us. While all our efforts seemed to have ended up in vain, we found that we had no choice but to start by taking a simplistic approach to reduce fairness to equality of agents' utility values (in some cases with a sufficiently small difference). Such an approach suffers from several drawbacks, as it is not able to capture many real-life situations. For example, in real life some people might consider it fair for rich people to receive less and poor people to receive more concerning the allocation of public resources. To this end, we looked into fairness models considering other types of fairness such as inequity-aversion and reciprocity. We hope the work presented in this book can shed light on this research area, which could attract more people to continue to work on it.

In this book we aim at tackling the issues of fairness and social optimality in multiagent systems from several aspects. We start with a relatively simple setting, in which agents use adaptive periodical strategy for coordination toward fairness. Agents interact with one another and review and adjust their strategies periodically. We then proceed to investigate how agents can coordinate to social optimality with reinforcement social learning in various types of games. After that, we discuss a related scenario of automated agent negotiation and present a negotiation strategy that aims at achieving a mutually beneficial agreement. We also considered a competitive negotiation problem and present an adaptive negotiation strategy enabling a rational agent to maximize his benefits from negotiation. Finally, we propose the novel idea of decision entrustment in two-agent repeated games—that is, an agent is allowed the option of entrusting his opponent to make joint decisions in bilateral interaction—and show that social optimality sustained by Nash equilibrium can be achieved.

There are several assumptions that we employ in the work presented in this book. The first one is that agents can always learn from their experiences, and the learning is always rational. That is to say, if an agent, through his experience, finds that an action is most likely to bring its best utility under certain conditions, then the agent should in principle use that action when conditions are met. This is most appropriate when the environment is cooperative. Second, agents should be able to observe the other agents—or some of them—in the system, including the other agents' actions and even the utilities they obtain through interaction. Last but not least, we base our work on some principles proposed by cognitive psychologists. For example, we use a theory by Fehr and Schmidt that people are inequality averse under certain conditions, as well as the research result due to Dawes and Thaleri that people are willing to be kind to those who are being kind and harsh to those who are being unkind. These research works provide a solid cognitive psychological basis for many of the assumptions used throughout this book.

Finally, we would like to express our gratitude for all the help from Higher Education Press and Springer in the publication of this book.

Tianjin, China — Jianye Hao
Hong Kong, China — Ho-fung Leung
August 2015

# Contents

# Chapter 1
# Introduction

Multiagent systems (MASs) have become a commonly adopted paradigm to model and solve real-world problems. Many competing definitions exist for a MAS. In this book, we consider a typical MAS as a system involving multiple autonomous software agents (or humans) interacting with each other with (possibly) conflicting interests and limited information, and the payoff of each agent is determined by the joint actions of all (or some) agents involved. Therefore, different from single-agent environments, in multiagent interaction environments, each agent needs to take other agents' behaviors into consideration when it makes its own decisions, since others' behaviors can directly influence what it expects from the system. The major question we seek to answer in this book can be summarized as follows: how can a *desirable goal* be achieved in different *multiagent interaction environments* where each agent may have its own limitations and (possibly) conflicting interests?

We divide different multiagent interaction environments into two major categories depending on the underlying intentional stance of the agents: cooperative multiagent environment and competitive multiagent environment. We say a multiagent interaction environment is cooperative if (most of) the agents within the system are willing to behave exactly as they are asked to even if this would cause conflicts with their individual interests. In other words, each agent is cooperative in achieving a commonly shared system-level goal even at the cost of its personal benefits. A multiagent interaction environment is considered to be competitive when each agent is only interested in maximizing its individual benefits. Thus, in competitive environments, agents may not have the incentive to follow what the system designer specifies and usually behave in purely individually rational manners.

We mainly focus on three major goals: fairness, social optimality, and individual rationality. To achieve either of the above goals, the research focus may vary and different approaches may be required depending on the nature of multiagent interaction problems being considered. In cooperative environments, since the agents are assumed to comply with the system designer's instructions, we usually need to consider how we can design an efficient learning strategy through which



J. Hao, H.-f. Leung, *Interactions in Multiagent Systems: Fairness, Social Optimality and Individual Rationality*, DOI 10.1007/978-3-662-49470-7_1

the agents can effectively coordinate their behaviors toward the goal we target at (e.g., fairness, social optimality) subject to practical limitations (e.g., limited information available, limited computational, and communication ability). On the contrary, in competitive environments, we mainly focus on addressing the following two questions:

- How can an agent obtain as much individual benefits as possible in the presence of other selfish agents?
- The second one is from the perspective of mechanism design: how can selfish agents be incentivized to autonomously adjust their behaviors toward the goal (e.g., social optimality) we target at?

To model the strategic interactions in MASs, the most powerful and commonly used mathematical tool is game theory. It provides a rich set of powerful and convenient mathematical frameworks to model different forms of strategic interactions among agents. For example, normal-form games can be used to model the single-shot strategic interactions among agents in which each agent makes decision simultaneously; for finitely/infinitely repeated games, which consists of finite/infinite repetition of stage games, they are naturally suitable for modeling the repeated (with unknown rounds) interactions among agents. Thus, in this book, we mainly study the problem of achieving different desirable goals by modeling agents' interactions as different types of games.

## 1.1 Overview of the Chapters

The book is organized in terms of the goals (fairness, social optimality, and individual rationality) we target at and the characteristic of the multiagent interaction environment (cooperative or competitive) we look at. Specifically, we focus on the following four parts, each of which corresponds to one chapter.

**Fairness in cooperative MASs** We pursue the goal of fairness in cooperative MASs from the following two perspectives. First, we describe an adaptive periodical strategy to enable the agents to coordinate toward fair and efficient outcomes in conflicting-interest games [1]. Second, two game-theoretic fairness models are described for two important multiagent interaction frameworks. The first is a game-theoretic fairness model for infinitely repeated game with conflict interest, in which it can be proved that there always exist fairness equilibria under which fair and optimal outcomes can be achieved [2]. The second is a game-theoretic fairness model for single-shot normal-form games, which can better explain actual human behaviors [3].

**Toward social optimality in cooperative MASs** In cooperative MASs, we describe two different types of learners for cooperative games, independent action learners and joint action learners, to coordinate on socially optimal outcomes under the social learning framework [4–6]. For general-sum games, we introduce a learning framework under which the agents can effectively learn to coordinate on socially optimal outcomes by adaptively choosing between individual learning and social learning [7]. Last, we turn to look at the practical bilateral negotiation problem and introduce an efficient negotiation framework for agents to achieve socially optimal allocations in cooperative bilateral negotiation environments [8].

**Maximizing individual benefits in competitive MASs** In competitive MASs, we focus on the bilateral negotiation setting in which two agents negotiate with each other with the goal of maximizing their individual utilities through negotiation. We describe an adaptive negotiation *ABiNeS* strategy which empirically shows to be very effective in competing with the state-of-the-art negotiation strategies [9, 10]. It is worth mentioning that the implementation of this *ABiNeS* strategy wins the champion of the third international Automated Negotiating Agents Competition [11] in both the qualifying and final rounds.

**Achieving social optimality in competitive MASs** In competitive MASs, we, as the system designer, may be also interested in achieving social optimality. However, in competitive environments, the agents are individually rational and interested in maximizing their own utilities only, which may have conflicts with the goal of social optimality. To resolve this conflict, we introduce an incentive mechanism based on sequential play to enable rational agents to have the motivation to adopt the policy of achieving socially optimal outcomes, which are successfully applied in two different learning frameworks [12–14].

## 1.2 Guide to the Book

The intended reader of this book can be a graduate student or an advanced undergraduate or any researcher in the areas of computer science, game theory, and any other related areas. The material of each chapter is independently organized and the readers can have no difficulty of understanding the materials by jumping to any interested chapter freely. We outline the chapters that follow.

- Chapter 2 first reviews some background concepts in game theory which will be used throughout the book and then gives an overall review of previous work in multiagent learning literature, which are divided into two major categories: cooperative multiagent systems and competitive multiagent systems. The previous works within each category are further divided according to the solution concepts they adopt, and how they link with the works in this book is also mentioned.
- Chapter 3 focuses on investigating the question of how to achieve fairness within cooperative multiagent environments. We interpret fairness from the following two perspectives. First, we focus on the multiagent interaction scenarios with conflict interest and our goal is to achieve fairness among agents in terms of equalizing the payoffs of agents as much as possible. Second, our human beings are not purely selfish but care about fairness, which cannot be explained and

predicted by the classical game theory. Thus, it is worthwhile to develop fairness-based game-theoretic model to better explain and predict human behaviors. Overall, we investigate the goal of fairness in MAS from two perspectives: strategy design and fairness game-theoretic model design, which are introduced in Sects. 3.1 and 3.2, respectively.

- Chapter 4 turns to the investigation of how to achieve social optimality within cooperative multiagent environments. Achieving social optimality means maximizing the sum of all agents' payoff involved in the interaction. In cooperative multiagent environments, the agents are assumed to be willing to cooperate toward social optimality, but the limited information available to each agent might impede effective coordination toward socially optimal outcomes among them. We model the agent interactions using two different types of games, cooperative games and general-sum games, and describe the corresponding state-of-the-art social learning frameworks toward social optimality in Sects. 4.1 and 4.2, respectively. Lastly we consider a particular multiagent negotiation problem and describe an efficient negotiation mechanism for achieving socially optimal allocations in Sect. 4.3.
- Chapter 5 focuses on the problem of maximizing an agent's individual benefits within competitive multiagent systems. In competitive multiagent systems, each agent is usually only interested in maximizing its individual benefits, and a natural question from a single agent's perspective is how to design an efficient strategy to obtain as much payoff as possible by exploiting its partners during interactions. Therefore, in this chapter, we consider the competitive bilateral negotiation environment and introduce a state-of-the-art efficient negotiation strategy to maximize the individual utility of a negotiating agent against other competitive opponents.
- Chapter 6 is devoted to investigating how to achieve social optimality (maximizing system level's performance) within competitive multiagent systems. As we mentioned before, in competitive multiagent systems, the agents may not have the incentive to cooperate toward the goal of social optimality which may conflict with its individual benefits. Therefore, from the system designer's perspective, we need to design efficient mechanism to incentivize selfish agents to cooperate toward socially optimal outcomes. Therefore, in this chapter, we introduce an interesting variation of sequential play and show how this can be applied in different competitive multiagent interaction environments to facilitate agents to achieve socially optimal outcomes without undermining their individual rationality and autonomy, namely, two-agent repeated interaction framework (Sect. 6.1) and social learning framework (Sect. 6.2).
- Chapter 7 concludes the book with a summary of the book and also points out some future research directions for the readers.

# References

1. Hao JY, Leung HF (2010) Strategy and fairness in repeated two-agent interaction. In: Proceedings of ICTAI'10, Arras, pp 3–6
2. Hao JY, Leung HF (2012) Incorporating fairness into infinitely repeated games with conflicting interests for conflicts elimination. In: Proceedings of ICTAI'12, Athens, pp 314–321
3. Hao JY, Leung HF (2012) Incorporating fairness into agent interactions modeled as two-player normal-form games. In: Proceedings of PRICAI'12, Kuching
4. Hao JY, Leung HF (2013) Reinforcement social learning of coordination in cooperative multi-agent systems(extended abstract). In: Proceedings of AAMAS'13, St. Paul, pp 1321–1322
5. Hao JY, Leung HF (2013) The dynamics of reinforcement social learning in cooperative multiagent systems. In: Proceedings of IJCAI'13, Beijing, pp 184–190
6. Hao JY, Leung HF, Ming Z (2014) Multiagent reinforcement social learning toward coordination in cooperative multiagent systems. ACM Trans Auton Adapt Syst 9(4):20
7. Hao JY, Leung HF (2011) Learning to achieve social rationality using tag mechanism in repeated interactions. In: Proceedings of ICTAI'11, Boca Raton, pp 148–155
8. Hao JY, Leung HF (2012) An efficient neogtiation protocol to achieve socially optimal allocation. In: PRIMA 2012: principles and practice of multi-agent systems, Kuching
9. Hao JY, Leung HF (2012) Abines: an adaptive bilateral negotiating strategy over multiple items. In: Proceedings of IAT'12, Macau, vol 2, pp 95–102
10. Hao JY, Song SZ, Leung HF, Ming Z (2014) An efficient and robust negotiating strategy in bilateral negotiations over multiple items. Eng Appl Artif Intell 34:45–57
11. Marsa-Maestre I, Lopez-Carmona MA, Ito T et al (2014) Novel insights in agent-based complex automated negotiation[M]. Springer, Tokyo
12. Hao JY, Leung HF (2012) Learning to achieve socially optimal solutions in general-sum games. In: PRICAI 2012: trends in artificial intelligence, Kuching, pp 88–99
13. Hao JY, Leung HF (2012) Achieving social optimality with influencer agents. In: Proceedings of complex'12, Santa Fe
14. Hao JY, Leung HF (2015) Introducing decision entrustment mechanism into repeated bilateral agent interactions to achieve social optimality. Auton Agents Multi-Agent Syst 29(4):658–682

## References

# Chapter 2
# Background and Previous Work

In this chapter, we first review some concepts and terminologies from game theory and multiagent systems areas that will be used throughout the book in Sect. 2.1. After that, we explore previous work in multiagent learning literature by dividing them into two major categories: cooperative multiagent systems and competitive multiagent systems, distinguished by the underlying intentional stance of the agents within the system. Within each category, we review previous work in three different parts distinguished by the different goal that the work targets at. In Sect. 2.2, we focus on reviewing previous work whose goal falls into one of the following major solution concepts: Nash equilibrium, fairness, or social optimality. In Sect. 2.3, we focus on investigating previous work targeting at either of the following major solution concepts: Nash equilibrium, maximizing individual benefits, and Pareto-optimality.

## 2.1 Background

Game theory has been the most commonly adopted mathematical tool for us to model the strategic interactions among different agents in different interaction scenarios. We usually model the one-time interaction among agents as single-shot normal-form game and finite/infinite repeated games to model the repeated interaction among agents.

### *2.1.1 Single-Shot Normal-Form Game*

In single-shot games, each agent is allowed to choose one action from its own action set simultaneously, and each agent will receive its own payoff based on the joint



J. Hao, H.-f. Leung, *Interactions in Multiagent Systems: Fairness, Social Optimality and Individual Rationality*, DOI 10.1007/978-3-662-49470-7_2

action of all agents involved in the interaction. Formally a $n$-player normal-form game $G$ is a tuple $\langle N, (A_i), (u_i)\rangle$ where

- $N = \{1, 2, \cdots, n\}$ is the set of players.
- $A_i$ is the set of actions available to player $i \in N$.
- $u_i$ is the utility function of each player $i \in N$, where $u_i(a_i, a_j)$ corresponds to the payoff player $i$ receives when the outcome $(a_i, a_j)$ is achieved.

We call each possible combination of all players' action (joint action) an outcome. There are a number of important solution concepts defined in normal-form games including Nash equilibrium, Pareto-optimality, and social optimality, which can be understood as a subset of outcomes satisfying special properties. These solution concepts also have been adopted as the learning goals when designing different learning strategies in multiagent learning area. In the following, we will introduce each of them in the context of two-player normal-form games which is the most commonly investigated case in the multiagent learning literature. The most important solution concept in game theory is Nash equilibrium. Under a Nash equilibrium, each agent is making its best response to the strategy of the other agent, and thus no agent has the incentive to unilaterally deviate from its current strategy.

**Definition 2.1** A *pure strategy Nash equilibrium* for a two-player normal-form game is a pair of strategies $(a_1^*, a_2^*)$ such that

1. $u_1(a_1^*, a_2^*) \geq u_1(a_1, a_2^*), \forall a_1 \in A_1$
2. $u_2(a_1^*, a_2^*) \geq u_2(a_1^*, a_2), \forall a_2 \in A_2$

For example, considering the two-player normal-form game in Fig. 2.1, we can easily check that there are two pure strategy Nash equilibria in this game: $(C, D)$ and $(D, C)$. Under each of these two outcomes, no agent has the incentive to unilaterally switch its current action to another one.

If the agents are allowed to use mixed strategy, then we can naturally define the concept of *mixed strategy Nash equilibrium* similarly.

**Fig. 2.1** An example of two-player conflicting-interest game

| 1's payoff, 2's payoff | | Player 2's action | |
|---|---|---|---|
| | | C | D |
| Player 1's action | C | 0, 0 | 1, 2 |
| | D | 2, 1 | 0, 0 |

**Definition 2.2** A *mixed strategy Nash equilibrium* for a two-player normal-form game is a pair of strategies $(\pi_1^*, \pi_2^*)$ such that

1. $\bar{U}_1(\pi_1^*, \pi_2^*) \geq \bar{U}_1(\pi_1, \pi_2^*), \forall \pi_1 \in \Pi(A_1)$
2. $\bar{U}_2(\pi_1^*, \pi_2^*) \geq \bar{U}_2(\pi_1^*, \pi_2), \forall \pi_2 \in \Pi(A_2)$

where $\bar{U}_i(\pi_1^*, \pi_2^*)$ is player $i$'s expected payoff under the strategy profile $(\pi_1^*, \pi_2^*)$, and $\Pi(A_i)$ is the set of probability distributions over player i's action space $A_i$. A mixed strategy Nash equilibrium $(\pi_1^*, \pi_2^*)$ is degenerated to a *pure strategy Nash equilibrium* if both $\pi_1^*$ and $\pi_2^*$ are pure strategies.

If we again use the same game in Fig. 2.1 as an example, we can verify that there also exists one mixed strategy Nash equilibrium: $\pi_i^*(C) = \frac{1}{3}$, $\pi_i^*(D) = \frac{2}{3}$, for $i = 1, 2$.

Another important solution concept in game theory is Pareto-optimality. An outcome is *Pareto-optimal* if and only if there does not exist another outcome under which no player's payoff is decreased and also at least one player's payoff is strictly increased. We can formalize it in the following definition:

**Definition 2.3** An outcome $s$ is *Pareto-optimal* if and only if there does not exist another outcome $s'$ such that $\forall i \in N$, $u_i(s') \geq u_i(s)$, and there exists some $j \in N$, for which $u_j(s') > u_j(s)$.

For example, considering the two-player normal-form game in Fig. 2.1, both the outcomes $(C, D)$ and $(D, C)$ are Pareto-optimal.

The last solution concept is social optimality, which refers to those outcomes under which the sum of all agents' payoffs is maximized. For example, considering the two-player normal-form game in Fig. 2.1, both the outcomes $(C, D)$ and $(D, C)$ are socially optimal, which coincide with the set of Pareto-optimal outcomes.

### *2.1.2 Repeated Games*

Repeated games can be used to model the repeated interactions among the same set of agents for finite or infinite number of times. In repeated games, usually a given normal-form game is played among the same set of players. If a strategic game is infinitely repeated, it is called an infinitely repeated game.

**Definition 2.4** An infinitely repeated game of $G$ is $\langle N, (A_i), (u_i), (\succsim_i) \rangle$ where

- $N$ is the set of $n$ players.
- $A_i$ is the set of actions available to player $i$.
- $u_i$ is the payoff function of each player $i$, where $u_i(a_i, \ldots, a_n) = p_i$, the payoff player $i$ receives when the outcome $(a_i, \ldots, a_n)$ is achieved. Here $a_i$ is the actions player $i$ chooses, and $(a_i, \ldots, a_n)$ is called an action profile.

- $\succsim_i$ is the preference relation for player $i$ which satisfies the following property: $O_1 \succsim_i O_2$ if and only if $\lim_{t\to\infty} \Sigma_{k=1}^{t}(p_1^k - p_2^k)/t \geq 0$, where $O_1 = (a_{i,t}^1, \ldots, a_{n,t}^1)_{t=1}^{\infty}$ and $O_2 = (a_{i,t}^2, \ldots, a_{n,t}^2)_{t=1}^{\infty}$ are the outcomes of the infinitely repeated game, and $p_1^k$ and $p_2^k$ are the corresponding payoffs player $i$ receives in round $k$ of outcomes $O_1$ and $O_2$, respectively.

The preference relation we adopt here is the limit of means preference relation which is the one we adopt in the book. Notice that there are also other different ways to define preference relation such as discounting and overtaking, and interested reader may refer to the book [1] for details. Also we usually assume that the action set of all agents are the same, i.e., $A_1 = \cdots = A_n = A$.

In infinitely repeated games, the strategy of an agent specifies the action choice of the agent for each round. Considering the large space of the agents' possible strategies, it is very difficult for us to identify all the Nash equilibria of a given infinitely repeated game. Thus, usually we turn to characterize the set of all possible payoff profiles that correspond to Nash equilibria of the infinitely repeated game. First, we need to introduce the concepts of *enforceable* and *feasible* payoff profiles which would be useful in the characterization of all possible Nash equilibrium payoff profiles of the infinitely repeated games.

**Definition 2.5** A payoff profile $(r_1, r_2)$ is enforceable if and only if $r_i \geq v_i$, $\forall i \in N$, where $v_i$ is agent $i$'s minimax payoff.

**Definition 2.6** A payoff profile $(r_1, r_2)$ is enforceable if and only if there exist nonnegative rational values $\alpha_a$ such that we have $\sum_{a\in A} \alpha_a u_i(a) = r_i$, $\forall i \in N$, and also $\sum_{a\in A} \alpha_a = 1$.

Based on the previous concepts, we can have the following theorem—folk theorem, which characterizes the set of all possible Nash equilibrium payoff profiles of the infinitely repeated games with limit of means criterion.

**Theorem 2.1** *If a payoff profile $r$ of game $G$ is both feasible and enforceable, then it is the payoff profile for some Nash equilibrium of the limit-of-means infinitely repeated game of $G$.*

## 2.2 Cooperative Multiagent Systems

### *2.2.1 Achieving Nash Equilibrium*

Convergence to Nash equilibrium has been the most commonly adopted goal to pursue within different multiagent environments in the multiagent learning literature. We review the representative work within this direction and also point out the limitations.

The interactions among agents are usually modeled as two-player repeated (or stochastic) games. In the work of Claus and Boutilier [2], two different types of learners are distinguished based on $Q$-learning algorithm, independent learner and joint-action learner; and their performance in the context of two-agent repeated cooperative games was investigated. An independent learner simply learns its $Q$-values for its individual actions by ignoring the existence of the other agent, while a joint-action learner learns the $Q$-values for the joint actions. Empirical results show that both types of learners can successfully coordinate on the optimal joint actions in simple cooperative games without significant performance difference. However, both of them fail to coordinate on optimal joint actions when the game structure becomes more complex, i.e., the climbing game (Fig. 4.1a) and the penalty game (Fig. 4.1b). For example, considering the penalty game in which $k = -100$, it has three Nash equilibria of which two of them ($(a, a)$ and $(c, b)$) are the optimal equilibria to achieve. Due to initial random explorations, both agents will find both actions $a$ and $c$ unattractive since mis-coordination on $(a, c)$ or $(c, a)$ results in significant payoff loss for both agents. In contrast choosing action $b$ is the more preferred option for both agents since nonnegative payoffs can always be guaranteed. After both agents have learned the policy of choosing action $b$, the outcome will be converged to the nonoptimal equilibrium $(b, b)$ due to the stable nature of equilibria.

A number of improved learning algorithms have been proposed afterward. Lauer and Rienmiller [3] propose the distributed $Q$-learning algorithm base on the optimistic assumption. Specifically, the agents' $Q$-values for each action are updated in such a way that only the maximum payoff received by performing this action is considered. Besides, the agents also need to follow an additional coordination mechanism to avoid mis-coordination on suboptimal joint actions. It is proved that optimal joint actions can be guaranteed to achieve if the cooperative game is deterministic; however, it fails when dealing with stochastic environments. The main difficulty dealing with stochastic domains can be intuitively explained as follows: in stochastic environments, apart from the effects of other agents' actions, the stochastic feature of the environment also influences the way of learning the optimal policy. However, the behaviors of the agents are expected to maximize based on the optimistic assumption, while the stochastic influence of the environment has to be taken into consideration with the expected values. When these two aspects of uncertainties are combined together, the agents may wrongly mix the estimated rewards of different actions, thus cannot learn the real maximal reward of each action.

Kapetanakis and Kudenko [4] propose the FMQ heuristic to alter the $Q$-value estimation function to handle the stochasticity of the games. Under FMQ heuristic, the original $Q$-value for each individual action is modified by incorporating the additional information of how frequent the action receives its corresponding maximum payoff. Experimental results show that FMQ agents can successfully coordinate on an optimal joint action in partially stochastic climbing games, but fail for fully stochastic climbing games. An improved version of FMQ (recursive FMQ) is proposed in [5]. The major difference with the original FMQ heuristic is that the $Q$-function of each action $Q(a)$ is updated using a linear interpolation based on the occurrence frequency of the maximum payoff by performing this action and bounded by the values of $Q(a)$ and $Q_{\max}(a)$. This improved version of FMQ is more robust and less sensitive to the parameter changes; however, it still cannot achieve satisfactory performance in fully stochastic cooperative games.

Matignon et al. [6] introduce the concept of hysteretic learning to facilitate better coordination on optimal Nash equilibrium in stochastic cooperative games. The general idea behind hysteretic learning is that agents should be cautious when they update their $Q$-values of their actions based on optimistic assumption. Under optimistic assumption, the $Q$-value of each action is updated only based on the maximal payoff obtained by taking the action, while all lower payoffs are considered as the results of other agents' random explorations and ignored. However, in stochastic environments, the lower payoffs of different actions may also be caused by the stochastic nature of the environment itself. To this end, under hysteretic learning, the authors propose that each agent should adopt two different learning rates to balance between the updates based on optimistic assumption and the update handling stochasticity of the environment. Simulation results show that hysteretic learners are able to successfully cover to optimal Nash equilibrium in partially stochastic climbing games over 80 % of the runs.

Panait et al. [7] propose the lenient multiagent learning algorithm to overcome the noise introduced by the stochasticity of the games. The basic idea is that at the beginning each lenient learner has high leniency (being optimistic), i.e., update the utility of each action based on the maximum payoff received from choosing this action and ignore those penalties occurred by choosing this action. Gradually, the learners will decrease their leniency degrees. Simulation results show that the agents can achieve coordination on the optimal joint action in the fully stochastic climbing game in more than 93.5 % of the runs compared with only around 40 % of the runs under FMQ heuristic.

Matignon et al. [8] review most of the existing independent multiagent reinforcement learning algorithms including decentralized $Q$-learning [2], distributed $Q$-learning [3], recursive FMQ [4], lenient $Q$-learning [7], and hysteretic $Q$-learning [6] in cooperative Markov games. The strength and weakness of these algorithms are evaluated and discussed in a number of multiagent domains including matrix games, Boutilier's coordination games, predator pursuit domains, and a special multistate game. Their evaluation results show that all of them fail to achieve coordination for fully stochastic games and only recursive FMQ can achieve coordination for 58 % of the runs.

To summarize, most previous work [2–5, 7, 9–11] studies the coordination problem within the framework of two players iteratively playing a single-stage cooperative game. However, little work has been done in an alternative important setting involving a large population of agents in which each agent interacts with another agent randomly chosen from the population each round. The interaction between each pair of agents is still modeled as a single-stage cooperative game, and the payoff each agent receives in each round only depends on the joint action of itself and its interacting partner. Under this framework, each agent learns its policy through repeated interactions with multiple agents, which is termed as *social learning* [12], in contrast to the case of learning from repeated interactions against the same opponent in the two-agent case. This social learning framework has been adopted to investigate the norm emergence problem in conflict-interest game (e.g., anti-coordination game) [12, 13], and it has been found that the agents can finally learn to follow an optimal norm through social learning. However, it is still not clear a priori whether the agents can learn to coordinate on an optimal joint action in cooperative games under such a social learning framework, which is the major question we are going to investigate in Sect. 4.1.

## 2.2.2 *Achieving Fairness*

### 2.2.2.1 Fairness in Behavioral Economics

The theory of inequity aversion has been used as one of the most popular theories for modeling fairness in behavioral economics [14] with the premise that people not only care about their own payoffs but also the relative payoffs with other people [15, 16]. Based on this theory, Fehr and Schmidt [15] propose an inequity aversion model to explain the phenomena that irrational actions instead of fully self-interested decisions can be made by people to prevent inequitable outcomes. This model is based on the assumption that people show a weak aversion toward advantageous inequity and a strong aversion toward disadvantageous inequity. In their fairness model, for a set of $n$ agents, the utility function of agent $i$ can be formally expressed as follows:

$$\begin{aligned} U_i(u) = u_i - \alpha_i \frac{1}{n-1} \sum_{j \neq i} \max\{u_j - u_i, 0\} - \\ \beta_i \frac{1}{n-1} \sum_{j \neq i} \max\{u_i - u_j, 0\}. \end{aligned} \tag{2.1}$$

Here $U_i(u)$ is the perceived utility of agent $i$, and $u = \{u_1, u_2, \ldots, u_n\}$ with $u_i$ the payoff received by agent $i$. The second term describes the utility cost when the payoff it receives is lower than that of other agents, and the third term represents the utility cost when the payoff it receives is higher than that of others. The parameters $\alpha_i$ and $\beta_i$ refer to the weighting factors representing the agent's suffering degree when it receives lower and higher payoff than others, respectively. Usually $\alpha_i \geq \beta_i$ is assumed in this model indicating that the agent suffers more from the inequity when it receives lower payoff than others. Moreover, $0 \leq \beta_i < 1$ is assumed here. By assuming $\beta_i \geq 0$, it is guaranteed that no agent is willing to receive higher payoff than others to increase inequity, and if $\beta_i \geq 1$, then the agent will be ready to throw away part of its own payoff to increase the equity among the agents. The authors have shown that this model can accurately predict human behavior in various games such as ultimatum game and public goods game with punishments.

Another line of research on modeling fairness in behavioral economics is based on the theory of reciprocity [17, 18]. Reciprocity means people tend to show kindness to those people that are kind to themselves and show unkindness to those people that are unkind to themselves. The key distinction with inequity-averse model is that this model also takes the kindness/unkindness intention of each action into account, and fairness is achieved by reciprocal behavior which is regarded as a response to kindness/unkindness instead of a desire for reducing inequity among payoffs.

Both types of models are designed targeting at analyzing single-shot simultaneous (or sequential) games such as ultimatum game and dictator game and have their own merits in explaining certain aspects of human behaviors that are inconsistent with the theory of individual rationality. However both types of models cannot be directly applied to repeated games. We believe that it is equally important to take into account fairness in the context of repeated game since it is a suitable framework for modeling multiagent repeated interactions. This issue will be further discussed and investigated in Chap. 3 (Sect. 3.2.2). Another limitation of previous work is that previous fairness models [15, 17] only take one fairness factor (either inequity aversion or reciprocity) into consideration, which thus only reflects a partial view of the motivations behind fairness. This issue will be further discussed and handled in Chap. 3 (Sect. 3.2.1).

#### 2.2.2.2 Achieving Fairness Through Multiagent Reinforcement Learning

A lot of approaches have been proposed for solving the multiagent learning problem based on reinforcement learning [19–22]. However, the criterion of fairness has not been given much attention in the literature of multiagent reinforcement learning, and we review two representative works in this direction.

Jong et al. [23] argue that it is important for a software agent to exhibit humanlike fairness behaviors since it is inevitable that an agent has to interact with human beings. To this end, they transform the descriptive inequity-averse fairness model from behavioral economic area into a computational model driven by continuous action learning automata for specifying agents' behaviors. A continuous action learning automata can be considered as a reinforcement learner with infinite actions. Each time learning automata select an action and receive a feedback from the environment and update the probability of choosing each action based on the feedback. In this work, the authors incorporate the inequity-averse fairness model into the decision-making process of the continuous action learning automata by mapping the original feedback from the environment to the utility value based on the inequity-averse fairness model each time. The authors apply this computational model into two representative game settings: the ultimatum game and Nash bargaining game. Simulation results show that the agents adopting this computational model can successfully exhibit fair behaviors that are better aligned with actual human behaviors observed in human experiments.

Nowé et al. [24, 25] firstly investigate the problem of how agents are able to coordinate on both fair and efficient outcomes through repeated interactions. To this end, they propose a periodical policy for multiple agents to achieve fair outcomes by incorporating the descriptive inequity-averse fairness model. Their policy involves two periods: reinforcement learning period and communication period. In the reinforcement learning period, the agents adopt $Q$-learning approach to choose actions with learning rate $\eta$. In the communication period, the agent who receives highest accumulative payoff and periodical payoff removes its current action out of its action space. The purpose is to give other agents the opportunity to reach their preferred outcomes. Simulation results show that each agent can obtain approximately equal average payoffs by adopting this periodical policy. However, this policy suffers from two major problems, i.e., stochastic exploration and non-robustness as follows, and this topic will be further investigated in Chap. 3 (Sect. 3.1).

**Stochastic exploration** Under the periodical policy, the agent whose performance is currently the best will exclude its current action from its action space, and then the agents will begin a new exploration phase using reinforcement learning method at the beginning of the next noncommunication period. The purpose of doing this is to give other agents the opportunity to achieve higher payoffs by achieving different Nash equilibria preferred by different agents. However, this also brings in the side effect that the agents will always explore their action space stochastically at the beginning of each noncommunication period, and thus it is inefficient in terms of the rate of convergence to another new Nash equilibrium. Additionally, it is possible that the agents will converge to an old Nash equilibrium which has been achieved before, and thus this will slow down the rate of achieving fairness between agents.

**Non-robustness** In the periodical policy, it is assumed that all agents obey the same coordination principle (i.e., excluding the current action if it is the best agent and reexploring its new action subspace at the beginning of each noncommunication period). However, this policy is vulnerable to exploitations of malicious agents when the system is open to outside agents, since the malicious agent can always choose its preferred action without following the coordination principle and its preferred Nash equilibrium will be achieved at the end of each noncommunication period.[1] In other words, the malicious agent always has the incentive to exploit others instead of obeying the coordination rule of the periodical policy. Therefore, fairness between agents cannot be guaranteed to be achieved when this kind of malicious agent exists.

[1] The rate of convergence to the Nash equilibrium preferred by the malicious agent depends on the learning rate of the agents, and this Nash equilibrium can always be achieved as long as the length of the noncommunication period is large enough.

### 2.2.3 Achieving Social Optimality

Learning to achieve coordination on (socially) optimal outcomes in the setting of a population of agents has received a lot of attention in multiagent learning literature [26–31]. Previous work mainly focuses on the coordination problem in two different games, prisoner's dilemma game (Fig. 2.2) and anti-coordination games (Fig. 2.3), which are representatives of two different types of coordination scenarios. For the first type of games, the agents need to coordinate on the outcomes with identical actions to achieve socially optimal outcomes, while in the second type of games, the achievement of socially optimal outcomes requires the agents to coordinate on outcomes with complementary actions. In previous work, tag has been extensively used as an effective mechanism to bias the interactions among agents in the population. Using tags to bias the interaction among agents in prisoner's dilemma game is first proposed in [32, 33], and a number of further investigations on the tag mechanism in one-shot or iterated prisoner's dilemma game are conducted in [34–36].

Hales and Edmonds [26] firstly introduce the tag mechanism originated in other fields such as artificial life and biological science into multiagent systems research to design effective interaction mechanism for autonomous agents. In their model for prisoner's dilemma game, each agent is represented by $L + 1$ bits. The first bit indicates the agent's strategy (i.e., playing $C$ or $D$), and the remaining $L$ bits are the tag bits, which are used for biasing the interaction among the agents and are assumed

**Fig. 2.2** An instance of the prisoner's dilemma game

| 1's payoff, 2's payoff | | Player 2's action | |
|---|---|---|---|
| | | C | D |
| Player 1's action | C | 3, 3 | 0, 5 |
| | D | 5, 0 | 1, 1 |

**Fig. 2.3** An instance of the anti-coordination game

| 1's payoff, 2's payoff | | Player 2's action | |
|---|---|---|---|
| | | C | D |
| Player 1's action | C | 0, 0 | 1, 2 |
| | D | 2, 1 | 0, 0 |

to be observable by all agents. In each generation, each agent is allowed to play the prisoner's dilemma game with another agent owning the same tag string. If an agent cannot find another agent with the same tag string, then it will play the prisoner's dilemma game with a randomly chosen agent in the population. The agents in the next generation are formed via fitness (i.e., the payoff received in the current generation) proportional reproduction scheme. Besides, low level of mutation is applied on both the agents' strategies and tags. This mechanism is demonstrated to be effective in promoting high level of cooperation among agents when the length of tag is large enough. They also apply the same technique to other more complex task domains which shares the same dilemma with prisoner's dilemma game (i.e., the socially rational action cannot be achieved if the agents are making purely individually rational decisions) and show that the tag mechanism can motivate the agents toward choosing socially rational behaviors instead of individually rational ones. However, there are some limitations of this tag mechanism. Since the agents mimic the strategy and tags of other more successful agents and the agents choose the interaction partner based on self-matching scheme, the agents can only play the same strategy with the interacting agents. Thus, socially optimal outcomes can be obtained only in the cases where the achievement of socially optimal outcomes requires the agents to coordinate on identical actions.

Chao et al. [30] propose a tag mechanism similar with the one proposed in [26], and the only difference is that mutation is applied on the agents' tags only. They evaluate the performance of their tag mechanism with a number of learning mechanisms commonly used in the field of multiagent system such as generalized tit-for-tat [30], WSLS (win-stay, loose-shift) [37], basic reinforcement learning algorithm [38], and evolutionary learning. They conduct their evaluations using two games: prisoner's dilemma game and coordination game. Simulation results show that their tag mechanism can have comparable or better performance than other learning algorithms they compared under both games. However, this tag mechanism suffers from similar limitations with Hales and Edmonds' model [26]: It only works when the agents are only needed to learn the identical strategies in order to achieve socially optimal outcomes, and this is indeed the case in both games they evaluated. It cannot handle the case when complementary strategies are needed to achieve coordination on socially optimal outcomes.

McDonald and Sen [27] reinvestigate the Hales and Edmonds' model [26] and the reason why large tags are required to sustain cooperation. The main explanation they come up with is that in order to keep high level of cooperation, sufficient number of cooperative groups is needed to prevent mutation from infecting most of the cooperative groups with defectors at the same time. There are two ways to guarantee the coexistence of sufficient number of cooperative groups, i.e., adopting high tag mutation rate or using large tags, which are all verified by simulation results. The limitation of Hales and Edmonds' model [26] is also pointed out in this paper, and they also provide an example (i.e., anti-coordination game in Fig. 2.3) in which Hales and Edmonds' model fails to promote the agents toward socially optimal behaviors. In anti-coordination game, complementary actions (either $(C, D)$ or $(D, C)$) are required for agents to coordinate on socially optimal outcomes.

However, under the self-matching tag mechanism in Hales and Edmonds' model, each agent only interacts with others with the same tag strings, and each interacting pair of agents usually shares the same strategy. Therefore, their mechanism only works when identical actions are needed to achieve socially optimal outcomes, while it fails when it comes to games like anti-coordination game.

Considering the limitation of Hales and Edmonds' model, McDonald and Sen [28, 29] propose three new tag mechanisms to tackle the limitation. The first tag mechanism is called tag-matching patterns (one and two sided). Each agent is equipped with both a tag and a tag-matching string, and agent $i$ is allowed to interact with agent $j$ if its tag-matching string matches agent $j$'s tag in one-sided matching. For two-sided matching, it also requires agent $j$'s tag-matching string to match agent $i$'s tag in order to allow the interaction between agent $i$ and $j$. By introducing this tag mechanism, the agents are allowed to play different strategy with their interacting partners. The second mechanism is payoff sharing mechanism, which requires each agent to share part of its payoff with its opponent. This mechanism is shown to be effective in promoting socially optimal outcomes in both prisoner's dilemma game and anti-coordination game; however, it can be applied only when side payment is allowed. The last mechanism they propose is called paired reproduction mechanism. It is a special reproduction mechanism which makes copies of matching pairs of individuals with mutation at corresponding place on the tag of one and the tag-matching string of the other at the same time. The purpose of this mechanism is to preserve the matching between this pair of agents after mutation in order to promote the survival rate of cooperators. Simulation results show that this mechanism can help in sustaining the percentage of agents coordinating on socially optimal outcomes at a high level in both the prisoner dilemma and anti-coordination games, and this mechanism can be applied in more general multiagent settings where payoff sharing is not allowed. However, similar to Hales and Edmonds' model [26], these mechanisms all heavily depend on mutation to sustain the diversity of groups in the system; thus, this leads to the undesired result that the variation of the percentage of coordination on socially optimal outcomes is very high all the time. Besides, the paired reproduction mechanism is shown to be effective in two types of symmetric games (prisoner's dilemma game and anti-coordination game) only, and it cannot be applied to the case of asymmetric games. The reason is that in asymmetric games, given a pair of interacting agents and their corresponding strategies, each agent may receive different payoffs depending on its current role (row or column agent). For example, considering the asymmetric game in Fig. 2.4, suppose a pair of interacting agents (agent $A$ and agent $B$) choosing actions $D$ and $C$, respectively. In this situation, agent $A$'s payoff is different when it acts as the row (payoff of 2) or column player (payoff of 4), even though both agents' strategies keep unchanged. However, in McDonald and Sen's learning framework, the role of each agent during interaction is not distinguished. They implicitly assume that all agents' roles are always the same in accordance with the definition of symmetric game. Thus, the payoffs of the interacting agents become nondeterministic when it comes to asymmetric games. Therefore, their learning framework is only applicable to those

**Fig. 2.4** Asymmetric anti-coordination game: the agents need to coordinate on the outcome (*D*, *C*) to achieve socially rational solution

| A's payoff, B's payoff | | Agent B's action | |
|---|---|---|---|
| | | C | D |
| Agent A's action | C | 0, 0 | 5, 4 |
| | D | 2, 15 | 0, 0 |

games that are agent symmetric (prisoner's dilemma game and anti-coordination game), i.e., the roles of the agents have no influence on their payoffs received.

To summarize, previous work [26, 28, 29] mainly focuses on modifying the interaction protocol from random interaction to tag-based interaction to evolve coordination on socially optimal outcomes. However, the agents' decision-making processes in the previous work are based on evolutionary learning (i.e., imitation and mutation) following evolutionary game theory, and thus the deviation of percentage of coordination is significant even though the average coordination rate is good. Another limitation of previous approaches is that it is explicitly assumed that each agent has access to all other agents' information (i.e., their payoffs and actions) in previous rounds, which may be unrealistic in many practical situations due to communication limitations. Finally, all previous work only focuses on two symmetric games, prisoner's dilemma game and anti-coordination game, and cannot be directly applied in the settings of asymmetric games. We will further discuss how to handle these limitations in Chap. 4 (Sect. 4.2).

## 2.3 Competitive Multiagent Systems

### *2.3.1 Achieving Nash Equilibrium*

Littman [39] considers the strictly competitive environment in which there are two agents (agent $i$ and $j$) and they have diametrically opposed goals. In this work, Markov game formalism is introduced as the mathematical framework for reasoning about the two-agent interaction environments. Markov games can be considered as the natural generalization of Markov decision process (MDP) and repeated game. Within this framework, the minimax-$Q$ learning algorithm is proposed by extending the original $Q$-learning algorithm in MDP to the framework of Markov games. The main difference between minimax-$Q$ learning and $Q$-learning is that the "max" operator used in the update step of $Q$-learning is replaced with the "minimax" operator, which is justified by the underlying assumption that the opponent agent $j$ always behaves to cause the greatest harm to agent $i$. The agents adopting minimax-$Q$ learning algorithm are theoretically guaranteed to converge to Nash equilibrium strategy profile in any two-player zero-sum Markov games under self-play, provided that every action and state are sampled infinitely often. However, when it comes to general-sum games, each agent may not be interested in minimizing its opponent(s)'s payoffs; thus, pursuing the solution concept of Nash equilibrium becomes less justified. An alternative learning goal is to maximize each agent's individual payoffs, which has been commonly adopted in the literature, and its related work will be reviewed next section.

### 2.3.2 Maximizing Individual Benefits

In competitive multiagent systems, agents usually aim at maximizing their individual benefits only. One of the most common and important practical multiagent competitive interaction scenarios is multiagent negotiation, which is our focus in this book. During negotiation, different negotiating parties usually have conflicting interests with each other and are only interested in obtaining as much utility as possible through negotiation. Until now great effects have been devoted to develop efficient negotiation strategies for automated negotiating agents in the literature. The most typical approach existing in the literature is concession-based strategy such as ABMP strategy [40]. The ABMP strategy agent decides on the next move based on its own utility space only and makes concession to its negotiating partners according to certain concession pattern. However, since the dynamics and influences of the negotiating partners are not taken into consideration, the weakness of this type of strategies is that it is difficult to reach those mutually beneficial outcomes and thus is inefficient in complex negotiation scenarios.

To overcome the aforementioned limitation, various techniques have been proposed to model certain aspects of the negotiation scenario to improve the efficiency of the negotiation outcome such as the opponent's preference profile and the knowledge of the negotiation domain [41–44]. Saha et al. [42] propose a learning mechanism using Chebychev's polynomials to approximately model the negotiating opponent's decision function in the context of repeated two-player negotiations. They prove that their algorithm is guaranteed to converge to the actual probability function of the negotiating partner under infinite sampling. Experiments also show that the agent using their learning mechanism can outperform other simple learning mechanisms and also be robust to noisy data. However, in their approach, it is assumed that the agents negotiate over one indivisible item (price) only, thus is not applicable to more general multi-issue negotiation scenarios.

Hindriks and Tykhonov [43] propose a Bayesian learning-based technique to model the negotiating opponent's private preference in the context of bilateral multi-issue negotiations. They test this technique on several negotiation domains and show that it can improve the efficiency of the bidding process by incorporating it into a negotiation strategy. One application of this modeling technique is that it is integrated into a negotiation strategy used by agent *TheNegotiator* [45] which participated in ANAC 2011 [45].

Brzostowski and Kowalczyk [46] propose a mechanism for predicting the negotiating partner's future behaviors based on the difference method. Based on the prediction results, the negotiation can be modeled as multi-stage control process, and the task of determining the optimal next-step offer is equivalent to the problem of determining the sequence of optimal control. Simulation results show that the agents using their mechanism can greatly outperform the classical approach in terms of utilities. However, their mechanism is only applicable in the single-item negotiation scenario. Besides, the underlying assumption of their mechanism is that the negotiation partner's strategy is the combination of time-dependent and behavior-dependent tactics, and thus may not work well against other types of negotiation partners.

To summarize, the major difficulty in designing automated negotiation agent is how to achieve optimal negotiation results given incomplete information on the negotiating partner. The negotiation partner usually keeps its negotiation strategy and its preference as its private information to avoid exploitations. A lot of research efforts have been devoted to better understand the negotiation partner by either estimating the negotiation partner's preference profile [43, 47, 48] or predicting its decision function [46, 49]. On one hand, with the aid of different preference profile modeling techniques, the negotiating agents can get a better understanding of their negotiating partners and thus increase their chances of reaching mutually beneficial negotiation outcomes. On the other hand, effective strategy prediction techniques enable the negotiating agents to maximally exploit their negotiating partners and thus receive as much benefit as possible from negotiation.

However, in most of previous work, the above two aspects are often investigated separately, and little efforts have been devoted to combine them together and evaluate the negotiation performance of various combinations of different techniques. To this end, in recent years a number of negotiation strategies, which take advantage of existing techniques from both aspects as previously mentioned, have been proposed, and agents employing these strategies have participated in *Automated Negotiating Agents Competition (ANAC)* [45, 50]. During the competitions, their performance has been extensively evaluated in a variety of multi-issue negotiation scenarios, and valuable insights have been obtained in terms of the advantages and disadvantages of different techniques, e.g., the efficacy of different acceptance conditions [44]. It is still an open and interesting problem to design more efficient automated negotiation strategies against a variety of negotiating opponents in different negotiation domains. We will further discuss this topic in Chap. 5.

### 2.3.3 Achieving Pareto-Optimality

Compared with the goal of Nash equilibrium and maximizing individual benefits, adopting Pareto-optimal solution has the advantage that it can prevent the agents from achieving loss-loss outcomes and also can improve all agents' utilities and the system level's utility most of the times. One well-known example is the prisoner's dilemma game (Fig. 2.2), in which if both agents only seek to maximize their individual payoffs, it will result in the inefficient Nash equilibrium (i.e., mutual defection). However, there exists a Pareto-optimal outcome under which both agents' utilities can be increased (mutual cooperation).

A number of approaches [51–53] have been proposed targeting at the prisoner's dilemma game only, and the learning goal is to achieve Pareto-optimal solution of mutual cooperation instead of Nash equilibrium solution of mutual defection. Besides, there also exists some work [54, 55] which addresses the problem of achieving Pareto-optimal solution in the context of general-sum games as follows. Sen et al. [54] propose an innovative expected utility probabilistic learning strategy by incorporating action revelation mechanism. Simulation results show that agents using action revelation strategy under self-play can achieve Pareto-optimal outcomes which dominate Nash equilibrium in certain games, and also the average performance with action revelation is significantly better than Nash equilibrium solution over a large number of randomly generated game matrices. Banerjee et al. [55] propose the conditional joint action learning (CJAL) strategy under which each agent takes into consideration the probability of an action taken by its opponent given its own action and utilizes this information to make its own decision. Simulation results show that agents adopting this strategy under self-play can learn to converge to the Pareto-optimal solution of mutual cooperation in prisoner's dilemma game when the game structure satisfies certain condition. However, all previous strategies are based on the assumption of self-play, and there is no guarantee of the performance against the opponents using other strategies.

In competitive environments, agents may not have the incentive to follow the same strategy specified by the system designer. To this end, a number of work [1, 56–58] assumes that the opponent may adopt different rational strategies and investigate the problem of how these agents can be incentivized to coordinate on Pareto-optimal outcomes. One representative work is folk theorem [1] in the literature of game theory. The basic idea of folk theorem is that there are some strategies based on punishment mechanism which can enforce desirable outcomes and are also in Nash equilibrium, assuming that all players are perfectly rational. From the multiagent learning perspective, the focus is to utilize the ideas in folk theorem to design efficient strategy against adaptive best-response opponents. Also the strategies we explore need not be in equilibrium in the strict sense, since it is very difficult to construct a strategy which is the best response to a particular learning strategy such as $Q$-learning. A number of teacher strategies [56, 57] have been proposed to induce better performance from the opponents via punishment mechanism, assuming that the opponents adopt best-response strategies such as $Q$-learning. Based on the teacher strategy, Goldfather++ [57], Crandall, and Goodrich [58] propose the strategy SPaM employing both teaching and following strategies and show its better performance in the context of two-player games against a number of best-response learners. The goal of the teaching component is to encourage its opponent to behave in the way desirable for the SPaM agent, and the goal of the following component is to guarantee that the SPaM agent can still achieve as high payoff as possible by exploiting its opponent while teaching its opponent at the same time.

Following this direction, we target at a more refined solution concept—socially optimal outcome, which represents those Pareto-optimal outcomes that are also socially optimal, i.e., maximizing the sum of all agents' payoffs involved. We investigate the problem of achieving socially optimal outcomes in competitive environments within two different types of learning contexts: two-player repeated interaction framework and a population of agents interacting with each other, which will be introduced in details in Chap. 6.

## References

1. Osborne MJ, Rubinstein A (1994) A course in game theory. MIT, Cambridge
2. Claus C, Boutilier C (1998) The dynamics of reinforcement learning in cooperative multiagent systems. In: Proceedings of AAAI'98, Madison, pp 746–752
3. Lauer M, Rienmiller M (2000) An algorithm for distributed reinforcement learning in cooperative multi-agent systems. In: Proceedings of ICML'00, Stanford, pp 535–542
4. Kapetanakis S, Kudenko D (2002) Reinforcement learning of coordination in cooperative multiagent systems. In: Proceedings of AAAI'02, Edmonton, pp 326–331
5. Matignon L, Laurent GJ, Le For-Piat N (2008) A study of FMQ heuristic in cooperative multi-agent games. In: AAMAS'08 workshop: MSDM, Estoril, pp 77–91
6. Matignon L, Laurent GJ, Le Fort-Piat N (2007) Hysteretic Q-learning: an algorithm for decentralized reinforcement learning in cooperative multi-agent teams. In: Proceedings of IEEE/RSJ international conference on intelligent robots and systems (IROS), San Diego, pp 64–69
7. Panait L, Sullivan K, Luke S (2006) Lenient learners in cooperative multiagent systems. In: Proceedings of AAMAS'06, Hakodate, pp 801–803
8. Matignon L, Laurent GJ, Le For-Piat N (2012) Independent reinforcement learners in cooperative markov games: a survey regarding coordination problems. Knowl Eng Rev 27:1–31
9. Panait L, Luke S (2005) Cooperative multi-agent learning: the state of the art. Auton Agents Multi-agent Syst 11(3):387–434
10. Wang X, Sandholm T (2002) Reinforcement learning to play an optimal nash equilibrium in team markov games. In: Proceedings of NIPS'02, Vancouver, pp 1571–1578
11. Brafman RI, Tennenholtz M (2004) Efficient learning equilibrium. Artif Intell 159:27–47
12. Sen S, Airiau S (2007) Emergence of norms through social learning. In: Proceedings of IJCAI'07, Hyderabad, pp 1507–1512
13. Villatoro D, Sabater-Mir J, Sen S (2011) Social instruments for robust convention emergence. In: Proceedings of IJCAI'11, Barcelona, pp 420–425
14. Camerer CF, Loewenstein G, Rabin M (2003) Advances in behavioral economics. Russell Sage Foundation Press/Princeton University Press, New York/Princeton
15. Fehr E, Schmidt KM (1999) A theory of fairness, competition and cooperation. Q J Econ 114:817–868
16. Bolton GE, Ockenfels A (2000) Erc-a theory of equity, reciprocity and competition. Am Econ Rev 90:166–193
17. Rabin M (1993) Incorporating fairness into game theory and economics. Am Econ Rev 83:1281–1302
18. Falka A, Fischbache U (2006) A theory of reciprocity. Games Econ Behav 54:293–315

19. Littman M (1994) Markov games as a framework for multi-agent reinforcement learning. In: Proceedings of ICML'94, New Brunswick, pp 322–328
20. Kapetanakis S, Kudenko D (2002) Reinforcement learning of coordination in cooperative multi-agent systems. In: Proceedings of AAAI'02, Edmonton, pp 326–331
21. Bowling M, Veloso M (2002) Multiagent learning using a variable learning rate. Artif Intell 136:215–250
22. Hu J, Wellman M (1998) Multiagent reinforcement learning: theoretical framework and an algorithm. In: Proceedings of ICML'98, Madison, pp 242–250
23. de Jong S, Tuyls K, Verbeeck K (2008) Artificial agents learning human fairness. In: Proceedings of AAMAS'08, Estoril. ACM, pp 863–870
24. Nowé A, Parent J, Verbeeck K (2001) Social agents playing a periodical policy. In: Proceedings of ECML'01, vol 2176. Springer, Berlin/New York, pp 382–393
25. Verbeeck K, Nowé A, Parent J, Tuyls K (2006) Exploring selfish reinforcement learning in repeated games with stochastic rewards. Auton Agents Multi-agent Syst 14:239–269
26. Hales D, Edmonds B (2003) Evolving social rationality for mas using "tags". In: Proceedings of AAMAS'03, pp 497–503. ACM, New York
27. Matlock M, Sen S (2005) The success and failure of tag-mediated evolution of cooperation. In: Proceedings of the first international workshop on learning and adaption in multi-agent systems. Springer, Berlin/New York, pp 155–164
28. Matlock M, Sen S (2007) Effective tag mechanisms for evolving coordination. In: Proceedings of AAMAS'07, Honolulu, p 251
29. Matlock M, Sen S (2009) Effective tag mechanisms for evolving coperation. In: Proceedings of AAMAS'09, Budapest, pp 489–496
30. Chao I, Ardaiz O, Sanguesa R (2008) Tag mechanisms evaluated for coordination in open multi-agent systems. In: Proceedings of 8th international workshop on engineering societies in the agents world, Athens, pp 254–269
31. Hao JY, Leung HF (2011) Learning to achieve social rationality using tag mechanism in repeated interactions. In: Proceedings of ICTAI'11, Boca Raton, pp 148–155
32. Holland JH, Holyoak K, Nisbett R, Thagard P (1986) Induction: processes of inferences, learning, and discovery. MIT, Cambridge
33. Allison PD (1992) The cultural evolution of beneficent norms. Soc Forces 71(2):279–301
34. Riol R, Cohen MD (2001) Cooperation withour reciprocity. Nature 414(6862):441–443
35. Hales D (2000) Cooperation without space or memory-tag, groups and the prisoner's dilemma. In: Multi-agent-based simulation, Boston
36. Howley E, O'Riordan C (2005) The emergence of cooperation among agents using simple fixed bias tagging. In: IEEE congress on evolutionary computation, Edinburgh
37. Nowak M, Sigmund K (1993) A strategy of winstay, lose-shit that outperforms tit-for-tat in the prisoner's dilemma game. Nature 364(6432):56–58
38. Wakano JY, Yamamura N (2001) A simple learning strategy that realizes robust cooperation better than pavlov in iterated prisoner's dilemma. J Ethol 19:9–15
39. Littman ML (1994) Markov games as a framework for multi-agent reinforcement learning. In: Proceedings of ICML'94, New Brunswick, pp 157–163
40. Jonker C, Robu V, Treur J (2006) An agent architecture for multi-attribute negotiation using incomplete preference information. Auton Agent Multi-agent Syst 15(2):221–252
41. Faratin P, Sierra C, Jennings NR (2003) Using similarity criteria to make negotiation trade-offs. Artif Intell 142(2):205–237
42. Saha S, Biswas A, Sen S (2005) Modeling opponent decision in repeated one-shot negotiations. In: Proceedings of AAMAS'05, Utrecht, pp 397–403
43. Hindriks K, Tykhonov D (2008) Opponent modeling in auomated multi-issue negotiation using Bayesian learning. In: Proceedings of AAMAS'08, Estoril, pp 331–338
44. Baarslag T, Hindriks K, Jonker C (2011) Acceptance conditions in automated negotiation. In: Proceedings of ACAN'11, Taibei

45. Baarslag T, Fujita K, Gerding EH, Hindriks K, Ito T, Jennings NR, Jonker C, Kraus S, Lin R, Robu V, Williams CR (2013) Evaluating practical negotiating agents: results and analysis of the 2011 international competition. Artif Intell 198:73–103
46. Jakub B, Ryszard K (2006) Predicting partner's behaviour in agent negotiation. In: Proceedings of AAMAS'06, Hakodate, pp 355–361
47. Zeng D, Sycara K (1998) Bayesian learning in negotiation. Int J Hum Comput Syst 48:125–141
48. Coehoorn RM, Jennings NR (2004) Learning an opponent's preferences to make effective multi-issue negotiation trade-offs. In: Proceedings of ICEC'04, Delft, pp 59–68
49. Zeng D, Sycara K (1996) Bayesian learning in negotiation. In: AAAI symposium on adaptation, co-evolution and learning in multiagent systems, Portland, pp 99–104
50. Baarslag T, Hindriks K, Jonker C, Kraus S, Lin R (2010) The first automated negotiating agents competition (ANAC 2010). In: Ito T, Zhang M, Robu V, Fatima S, Matsuo T (eds) New trends in agent-based complex automated negotiations. Springer, Berlin/Heidelberg, pp 113–135
51. Stimpson JL, Goodrich MA, Walters LC (2001) Satisficing and learning cooperation in the prisoner's dilemma. In: Proceeding of IJCAI'01, Seattle, pp 535–540
52. Moriyama K (2007) Utility based Q-learning to maintain cooperation in prisoner's dilemma game. In: Proceedings of IAT'07, Silicon Valley, pp 146–152
53. Moriyama K (2008) Learning-rate adjusting Q-learning for prisoner's dilemma games. In: Proceedings of WI-IAT'08, Sydney, pp 322–325
54. Sen S, Airiau S, Mukherjee R (2003) Towards a pareto-optimal solution in general-sum games. In: Proceedings of AAMAS'03, Melbourne, pp 153–160
55. Banerjee D, Sen S (2007) Reaching pareto optimality in prisoner's dilemma using conditional joint action learning. In: Proceedings of AAMAS'07, Honolulu, pp 211–218
56. Littman ML, Stone P (2001) Leading best-response strategies in repeated games. In: IJCAI workshop on economic agents, models, and mechanisms, Seattle
57. Littman ML, Stone P (2005) A polynomial time nash equilibrium algorithm for repeated games. Decis Support Syst 39:55–66
58. Crandall JW, Goodrich MA (2005) Learning to teach and follow in repeated games. In: AAAI workshop on multiagent learning, Pittsburgh

# Chapter 3
# Fairness in Cooperative Multiagent Systems

In cooperative MASs, the interests of individual agents are usually consistent with that of the overall system. In the cases of conflicting interest, each agent is assumed to have the willingness to cooperate toward a common goal of the system even at the cost of sacrificing its own benefits. Therefore, the behaviors of each agent in cooperative environments can be determined by the designer(s) of the system, which thus allows for intricate coordination strategies to be implemented beforehand. To achieve effective coordinations among agents, one traditional approach is to employ a superagent to determine the behaviors for all other agents in the system. However, there exist a number of disadvantages by adopting this approach. First, the scalability problem will become serious when the number of agents is significantly increased, since the computational space of the superagent increases exponentially to the number of agents. Second, it explicitly requires the superagent to be able to communicate with all agents in the system and has the global information, which may not be possible in distributed environments, where the communication cost can be very high. Lastly, it makes the system very vulnerable since the malfunction of the superagent would lead to the failure of the whole system.

To handle the above problems, an alternative approach, *current learning* [1], is to distribute the decision-making power into the hands of each individual agent in the system and each agent is responsible for determining its own behaviors. One central research question in this direction is to consider how individual agents can learn effectively to coordinate their behaviors to achieve some particular optimal solution based on their local information and limited communication capabilities.

In this chapter, we focus on the solution of fairness, which is an important and desirable goal to pursue in many multiagent applications such as resource allocation or job scheduling problem. In these practical problems, it is important to guarantee fairness among agents when the global performance of the system is determined by the agent with the worst performance. In this chapter, we approach the goal of fairness from the following two perspectives.



J. Hao, H.-f. Leung, *Interactions in Multiagent Systems: Fairness, Social Optimality and Individual Rationality*, DOI 10.1007/978-3-662-49470-7_3

Firstly, from concurrent learning perspective, we are interested in how the agents can effectively learn to coordinate toward the goal of fairness with limited communication capability. Specifically we introduce a state-of-the-art adaptive periodical strategy which enables the agents to coordinate toward fair and efficient outcomes in conflicting-interest games [2].

Secondly, we approach the goal of fairness in multiagent systems from game-theoretic perspective. Game theory is a commonly adopted mathematical framework for modeling the interaction among agents, and the commonly adopted assumption is individual rationality, i.e., each agent is assumed to be purely selfish. However, in real life people also care about fairness: the relative payoffs between them also matter. In MASs, it is usually inevitable for a software agent to interact with humans. To better understand human behaviors and also design strategies that can have better alignments with human behaviors, it can be useful to have some game-theoretic models which incorporate the fairness motive explicitly. Thus our goal is to better understand human fairness behaviors using game-theoretic analysis. In this chapter, we introduce two state-of-the-art game-theoretic fairness models for two different multiagent interaction frameworks. The first one is a game-theoretic fairness model for infinitely repeated game with conflict interest [3] (Sect. 3.1). The second one is a game-theoretic fairness model for single-shot normal-form games to better explain human behaviors [4] (Sect. 3.2).

## 3.1 An Adaptive Periodic Strategy for Achieving Fairness

### *3.1.1 Motivation*

In MASs, an important ability of an agent is to be able to coordinate effectively with other agents toward desirable outcomes. In multiagent environments, the stationary assumption of the environment is violated since for each agent the outcome not only depends on the action the agent itself takes but also the actions taken by other agents coexisting in the environment. Accordingly, the theoretical guarantees of convergence of the reinforcement learning techniques (e.g., Q-learning [5]) in single-agent environment do not hold any more when it comes to a multiagent interacting environment. Until now, there have been various multiagent reinforcement learning algorithms proposed with some success through applying single-agent reinforcement learning techniques to multiagent interacting environments with some success. For example, Claus and Boutilier [6] introduce two types of learners, independent learners (ILs) and joint action learners (JALs), which are natural extensions of single-agent Q-learning algorithm to multiagent cooperative environments, and the theoretical property of convergence for both ILs and JALs to one (deterministic) Nash equilibrium is also given. Another representative work is minimax Q-learning algorithm [7], which has the nice property of converging to a Nash equilibrium in zero-sum games under self-play, by modifying the single-agent

**Fig. 3.1** An example of two-player conflicting-interest game

| 1's payoff, 2's payoff | | Player 2's action | |
|---|---|---|---|
| | | C | D |
| Player 1's action | C | 0, 0 | 1, 2 |
| | D | 2, 1 | 0, 0 |

Q-learning technique based on the minimax solution concept borrowed from game theory.

While the property of convergence is desirable in terms of ensuring the stability of the system, convergence to Nash equilibrium may not always be the best choice. Let us consider a two-player game with conflicting interests shown in Fig. 3.1. Converging to one of the pure strategy Nash equilibria $(C, D)$ or $(D, C)$ can violate fairness criterion, i.e., it will be always unfair for one of the players if any of the pure strategy Nash equilibria is played, due to the unequal payoffs obtained by the two players. As an alternative, the agents can also learn to play the mixed strategy Nash equilibrium, in which the agents can receive equal payoffs (i.e., fairness is satisfied.). However, the average payoff each agent can receive is only $\frac{2}{3}$, which is much lower than the minimum payoff when any pure strategy Nash equilibrium is achieved due to high frequency of conflicting outcomes under the mixed strategy Nash equilibrium. To achieve both equal and maximum payoffs for both agents, the agents should be able to learn to coordinate on both pure strategy Nash equilibria $(C, D)$ and $(D, C)$ with equal chance.

To achieve coordination on fair outcomes, previous work [8, 9] proposes incorporating the periodical action removal mechanism into a reinforcement learning algorithm. However, it suffers from two major drawbacks, i.e., stochastic exploration and non-robustness. To tackle these problems, we introduce a novel adaptive strategy which enables agents to coordinate their action selection process more efficiently. Additionally, this strategy is robust to malicious agents. Different from previous works based on reinforcement learning approach, the novelty of this strategy is that it enables each agent to coordinate its action selection efficiently with its opponent by adjusting its attitudes in an adaptive way which is inspired from human decision-making theories [10–12]. In this way, the agents' action selection can be more well targeted with less stochastic exploration, and thus the number of times coordinating on conflicting outcomes can be greatly reduced. By comparing with previous work, it can be shown that the agents adopting this strategy are able to achieve fair outcomes with less payoff cost. There also exist a number of desired theoretical properties for this strategy, which are also validated in simulation. Additionally, by enabling adaptive adjustment of the adjustment rate, this adaptive strategy is robust to the exploitation of malicious agents, which is an important property in an open system.

## 3.1.2 *Problem Specification*

### 3.1.2.1 A Scenario of Repeated Agent Interaction

In distributed resource allocation or job scheduling problem, the agents often have to learn to coordinate their action selections repeatedly based on their local information in order to achieve fair allocation of resources for each agent or job. This kind of situation can be modeled by a repeated game with conflicting interests, in which multiple Nash equilibria exist and each agent prefers one of the Nash equilibria different from other agents'. Here we assume that an agent can receive higher payoff than its opponents when playing the Nash equilibrium it prefers. If an agent always receives lower payoff than its opponents no matter which Nash equilibrium is played, there is no way to achieve fairness between the agents and thus it is not considered. The outcomes will be unfair for at least one agent in the system if any pure strategy Nash equilibrium is played since the Nash equilibrium is not preferred by those agents which receive lower payoff than other agents.

A two-player matrix game with conflicting interests has already been shown in Fig. 3.1, in which there exist two pure strategy Nash equilibria $(R1, R2)$ and $(R2, R1)$. In this game, the player in the row prefers the Nash equilibrium $(R1, R2)$, and the player in the column prefers the Nash equilibrium $(R2, R1)$; thus it is unfair for one of the players if they play any single pure strategy Nash equilibrium in the game. Formally, the game with conflicting interests can be described by a tuple $\langle N, (A_i), (u_i)\rangle$ where $N = \{1, 2\}$ is the set of players, $A_i = \{R1, R2\}$ is the set of actions available to player $i$, $R1$ and $R2$ refer to the actions available to player $i$, and $u_i : A_1 \times A_2 \rightarrow \Re$ is the payoff function of each player $i$. The payoff matrix for the players is represented in Fig. 3.2, e.g., for $(u_1^1, u_2^1)$, $u_1^1$ refers to the payoff player 1 receives and $u_2^1$ refers to the payoff player 2 receives when both of the players play action $R1$ simultaneously. Without loss of generality, we assume that there exist two pure strategy Nash equilibria $(u_1^2, u_2^2)$ and $(u_1^3, u_2^3)$ in the conflicting-interest game, and conflicting interest means that player 1 prefers the equilibrium $(u_1^2, u_2^2)$ and player 2 prefers the equilibrium $(u_1^3, u_2^3)$. In general, the following inequalities in payoffs hold for player 1 $u_1^2 > u_1^3, u_1^3 > u_1^1, u_1^2 > u_1^4$ and for player 2 $u_2^3 > u_2^2, u_2^2 > u_2^1, u_2^3 > u_2^4$.

**Fig. 3.2** Payoff matrix for two players' conflicting-interest game

Player 2

R1 R2

Player 1

$$\begin{array}{c} R1 \\ R2 \end{array} \begin{pmatrix} (u_1^1, u_2^1) & (u_1^2, u_2^2) \\ (u_1^3, u_2^3) & (u_1^4, u_2^4) \end{pmatrix}$$

We assume that the agents take their actions independently and each agent only has access to its own payoff information without awareness of the opponent's information (e.g., its action and payoff); thus the problem described here can be generalized to situations where the opponent's information is complex or unknown. Besides, the communication cost between the agents is taken into account since the communication cost can be extremely high in an open, distributed environment. Overall the key issue is how the agents can learn to achieve fair outcomes (i.e., the equal payoffs each agent receives) in an efficient way (i.e., obtaining as high payoffs as possible). At the same time the communication cost between them should be reduced as much as possible.

#### 3.1.2.2 Fairness in Repeated Two-Agent Interaction

For better analyzing the outcomes of the repeated game, we give the following definitions.

**Definition 3.1** The *accumulated payoff* $CU_i(t)$ agent $i$ receives by time step $t$ is the sum of the payoffs agent $i$ receives in the discrete time interval $[0, t]$.

It is worth pointing out that there are many ways to achieve fairness, e.g., in Fig. 3.1, fairness can be achieved if only the conflicting outcome $(R1, R1)$ is achieved infinitely. Here we are interested in the case when the agents coordinate their actions such that the two pure strategy Nash equilibria are achieved alternately, since the agents can not only achieve fairness but also obtain as high payoff as possible in this situation. Therefore the definition of fairness is given as follows:

**Definition 3.2** At any time step $t$, *fairness* between the agents is achieved if and only if each agent in the repeated game receives the same accumulated payoffs by time step $t$ with the premise that each of the two pure strategy Nash equilibria is achieved for a finite number of rounds alternately.

In certain repeated game with conflicting interests, it is impossible to achieve above fairness between the agents since the agents are not able to obtain exactly the same accumulated payoffs due to the payoff structure of the game. Therefore we have the following definition of fairness in a more general context.

**Definition 3.3** At any time step $t$, $\epsilon$*-fairness* between the agents is achieved if and only if the following condition is satisfied at time step $t$: $|CU_1(t) - CU_2(t)| \leq \epsilon$ with the premise that the two pure strategy Nash equilibria are achieved alternately, where $CU_1(t)$ and $CU_2(t)$ are the accumulated payoffs agent 1 and agent 2 receive by time step $t$ respectively.[1]

Several notions of social welfare (e.g., egalitarian social welfare and envy-freeness) have been suggested to reflect fairness in social choice theory [13]. To

[1] $\epsilon$ is assumed to be a positive real number.

measure fairness in repeated games with conflicting interests in a more quantitative way, we use the following criteria to evaluate the game performance in fairness-related aspects.

*Time to fairness* Time to fairness (*TTF*) is defined as the length of time from the beginning of the game to the first time that fairness between the agents is achieved. Formally, *TTF* is the minimum of time $t$ which satisfies the following condition: $CU_i(t) = CU_j(t)$ with the premise that the two pure strategy Nash equilibria are achieved alternately, where $CU_i(t)$ and $CU_j(t)$ are the accumulated payoffs of agent $i$ and agent $j$ by time step $t$, respectively.

*Utilitarian social welfare* The utilitarian social welfare is defined as the sum of individual payoffs received by the agents of the system [13], and it provides a suitable metric for evaluating the overall performance of the system. Though the primary goal of the adaptive strategy is to achieve fairness in the repeated game, we have to guarantee that the utilitarian social welfare in the system is as high as possible, and also here we use each agent's average accumulated payoff over time as an alternative to evaluate this metric.

*Length of unfairness* This criterion is defined as the minimum length of time interval between two time steps that fairness is achieved in the agents. Formally, length of unfairness (LOU) can be represented by the following formula: LOU $=$ $\min\{t_2 - t_1\}$ where $t_1$ and $t_2$ ($t_2 > t_1$) are the time steps which satisfy the condition that $CU_i(t_1) = CU_j(t_1)$ and $CU_i(t_2) = CU_j(t_2)$.

### 3.1.3 An Adaptive Periodic Strategy

In this strategy, each agent chooses actions according to the attractiveness of each action, which is determined by two basic factors: non-conflict ratio and preference. According to prospect theory [11], different people have different risk attitudes toward these two factors when making their decisions. Inspired by the human decision-making theory, for each agent, we also associate an attitude $a_r^t \in [0, 1]$ with each action $r$ at time $t$, which is used as a weighting factor between the two basic factors. Therefore the value of the attractiveness of each action can be calculated as the weighted average of non-conflict ratio and preference. If $a_r^t = 0$, then the agent makes the decision based on the non-conflict ratio only; if $a_r^t = 1$, then the agent is most preference-relying indicating that it makes the decision based on preference for each action only. Formally, the attractiveness $\text{attr}_r^t$ for each action $r$ at time $t$ is calculated as following:

$$\text{attr}_r^t = (1 - a_r^t) \times h_r^t + a_r^t \times u_r \tag{3.1}$$

where $h_r^t$ is the non-conflict ratio of action $r$ at time $t$ and $u_r$ is the preference for action $r$.

The non-conflict ratio of each action keeps the record of the number of times that both agents take different actions over the length of history $h_l$, and we call it conflict occurring on an action if both agents take the same action simultaneously. If an agent makes decisions based on the non-conflict ratio only expecting for minimizing the possibility of conflict in the future, it will take the action which causes the least number of conflicts in the past. The non-conflict ratio $h_r^t$ can be expressed as follows:

$$h_r^t = \frac{n_r^t}{h_l} \tag{3.2}$$

where $h_l$ is the length of history the agent keeps record of and $n_r^t$ is the number of times that no conflict occurs on action $r$ within the history of length $h_l$ at time $t$.

The preference that each agent holds for each action represents the net payoff $u_r$ the agent can receive by taking the action $r$ if no conflict occurs when the agent takes this action.[2] If an agent only considers the preference when making its decision expecting for obtaining highest payoffs without conflict, it will take the action with potentially highest payoff.

Take the two player's conflicting-interest game in Fig. 3.1 as an example; for each agent, it takes action $R1$ at time $t$ if $\text{attr}_{R1}^t > \text{attr}_{R2}^t$, takes action $R2$ if $\text{attr}_{R1}^t < \text{attr}_{R2}^t$, and chooses them randomly if $\text{attr}_{R1}^t = \text{attr}_{R2}^t$. From the theory of conditions of learning in psychology [12], we know that attitude will be changed by favorable or unfavorable experience. Inspired from it, the attitude of agents also can be changed adaptively by the adjustment rate $a_\Delta$. At the end of each time step $t$, each agent updates its attitude $a_r^t$ accordingly: if the agent takes action $r$ and no conflict occurs, then its attitude is increased by $a_\Delta$, and if the agent takes action $r$ and conflict occurs, its attitude is decreased by $a_\Delta$.

It can be proved that only one of the pure strategy Nash equilibria will be converged to if the agents adopt above adaptive strategy shown in Theorem 3.4. Note that which Nash equilibrium is reached depends on the relative value of the agents' attitudes. However, the remaining problem is that reaching either of the pure strategy Nash equilibria is unfair for one of the agents, i.e., it always receives less payoffs than another agent. For achieving fairness between the agents, additional mechanism is needed to adjust the agents' attitudes based on their payoff information. In behavioral economics, a number of descriptive fairness models have been proposed to explain the fairness phenomenon among humans. In this strategy the descriptive fairness model for describing humans is incorporated into designing agents' strategy to achieve fairness among agents. Specifically the descriptive fairness model adopted here is the inequity aversion-based fairness model [10]. Each agent can be regarded as an inequity-averse agent, which adjusts its attitude

[2]The implicit assumption here is that each agent knows the potentially maximum payoff by taking each action. We claim that this assumption is reasonable since the agents can have access to this information in advance in practical application such as resource allocation problem with conflicting interests, which can be naturally modeled by conflicting-interest games.

autonomously based on the payoff information and perceived utility in order to achieve fairness in the system. Since we are only considering the case with two agents here, the aforementioned utility function in this fairness model can be simply reduced to the following:

$$U_i(u) = u_i - \alpha_i \max\{u_j - u_i, 0\} - \beta_i \max\{u_i - u_j, 0\}, i \neq j. \tag{3.3}$$

The details of the adjustment mechanism are specified as follows: according to utility function (3.3), each agent needs its own payoff information and the other agent's payoff information to calculate the perceived utility. Since it is assumed that each agent only knows its own information including its own action and the payoff it receives, the payoff information of each agent including the periodical payoff $u_i$ (the sum of the payoffs received within the current period) and the accumulated payoff $CU_i$ has to be publicized to the other agent in the system at the end of each period with periodical length $T$.[3] Since the payoff information publicizing is done only at the end of each period, the perceived utility of each agent can be obtained with least communication. After this, each agent updates its own attitudes accordingly at the end of each period. The updating procedure is specified by the following rules: if agent 1's accumulated payoff $CU_1$ and periodical payoff $u_1$ are both higher than the agent 2's at the end of each period, agent 1 decreases its attitudes by $\Delta U_- = \beta_1 \frac{(u_1-u_2)}{u_1}$, and agent 2 increases its attitudes by $\Delta U_+ = \alpha_2 \frac{(u_1-u_2)}{u_2}$. By updating their attitudes in this way, agent 2 can take its initiative to select its preferred action when it is in an unfair position, and also agent 1 will be more willing to give up its current action to coordinate on the Nash equilibrium agent 2 prefers. If agent 1's accumulated payoff $CU_1$ and periodical payoff $u_1$ are both lower than the agent 2's at the end of each period, agent 1 increases its attitudes by $\Delta U_+ = \alpha_1 \frac{(u_2-u_1)}{u_1}$, and agent 2 decreases its attitudes by $\Delta U_- = \beta_2 \frac{(u_2-u_1)}{u_2}$. The rationale for this updating rule is similar with the previous one and we omit it here. Here $\alpha_i$ and $\beta_i$ are the weighting factors of agent $i$ in the inequity aversion model, and the values of $\Delta U_-$ and $\Delta U_+$ for agent $i$ are the ratio of the absolute difference between agent $i$'s periodical payoff and its perceived utility to its periodical payoff times the weighting factor $\beta_i$ and $\alpha_i$, respectively.

By updating each agent's attitudes following previous rules, we are expecting that the agents can coordinate their actions efficiently so that the two pure strategy Nash equilibria are achieved alternately, and thus fairness can be achieved between the agents as well. Additionally, each agent decreases its adjustment rate $a_\Delta$ to zero if the number of periods that its periodical payoff or accumulated payoff is lower than another agent is above a threshold $m$.[4] By adding this mechanism, the agents

[3] Here we assume that the agents are always truthfully revealing their payoff information. There are various mechanisms (e.g., adopt a truthful third party to supervise this process) to guarantee this, which is beyond the scope of the book.

[4] The value of $m$ reflects the tolerance degree of the agents to malicious agents and thus can be set to different values accordingly.

can be prevented from being exploited by malicious agents that do not follow this adaptive strategy more than $m$ periods (e.g., always choose the action they prefer). In other words, this mechanism guarantees that the agent will always choose the action it prefers after it receives lower payoff than its opponent for consecutive $m$ periods. Therefore the agent adopting the adaptive strategy can be resistant to malicious exploitation of its opponent which does not follow this strategy. This mechanism can be regarded as a form of trigger strategy so that by implementing it the agents will have no incentive to deviate from this adaptive strategy. This mechanism is optional and can be activated when the agents are situated in an open environment.[5] The detailed description of this adjustment strategy is given in Algorithm 1.

Let's take the two player's conflicting-interest game in Fig. 3.1 as an example to further illustrate this adjustment strategy. Suppose that the outcome converges to $(R2, R1)$ at the end of period $t$, and agent 1 receives lower accumulated payoff than agent 2. For achieving fairness, agent 1's desired outcome $(R1, R2)$ has to be achieved later. In the adjustment strategy, agent 1 increases both of its attitudes by $\Delta U_{+}$; thus the difference between its values of $\text{attr}_{R1}$ and $\text{attr}_{R2}$ will be decreased.[6]

**Algorithm 1** Fairness adjustment strategy for agent $i$

**for** each time interval of length $T$ **do**
  broadcast its periodical payoff $u_i$ and accumulated payoff $CU_i$ to another agent $j$ in the system
  receive another agent $j$'s periodical payoff $u_j$ and accumulated payoff $CU_j$
  calculate its perceived utility $U_i$ according to utility function (3.3)
  **if** $u_i > u_j$ and $CU_i > CU_j$ **then**
    calculate $\Delta U_{-} = \frac{(u_i - U_i)}{u_i} = \beta_i \frac{(u_i - u_j)}{u_i}$
    **for** each attitude $a_r^t$ of agent $i$ on action $r$ at current time step $t$ **do**
      $a_r^{t+1} = a_r^t - \Delta U_{-}$
    **end for**
  **else if** $u_i < u_j$ and $CU_i < CU_j$ **then**
    calculate $\Delta U_{+} = \frac{(u_i - U_i)}{u_i} = \alpha_i \frac{(u_j - u_i)}{u_i}$
    **for** each attitude $a_r^t$ of agent $i$ on action $r$ at current time step $t$ **do**
      $a_r^{t+1} = a_r^t + \Delta U_{+}$
    **end for**
  **else**
    do nothing
  **end if**
  **if** $u_i < u_j$ or $CU_i < CU_j$ for consecutive $m$ periods **then**
    set its adjustment rate $a_{\Delta}$ to zero
  **end if**
**end for**

[5]The following analysis assumes that both agents in the system adopt this adaptive strategy and thus this mechanism is not activated.

[6]The reason behind is stated as follows: since agent 1 chooses action $R2$ at the end of period $t$, $\text{attr}_{R1} < \text{attr}_{R2}$ is satisfied at this time, and the difference between $\text{attr}_{R1}$ and $\text{attr}_{R2}$ can be expressed as $\text{diff} = \text{attr}_{R2} - \text{attr}_{R1}$. After agent 1 increases both of its attitudes by $\Delta U_{+}$, the difference

If $\Delta U_+$ is too small, it is possible that $\text{attr}_{R1}$ is still smaller than $\text{attr}_{R2}$; thus agent 1 will not change its action, and the outcome will still be $(R2, R1)$. If $\Delta U_+$ is large enough such that $\text{attr}_{R1} > \text{attr}_{R2}$ is satisfied for agent 1, agent 1 will start to choose action $R1$. If $\Delta U_+$ becomes larger, agent 1 will stick to action $R1$ for more time steps. If agent 1 still sticks to action $R1$ when agent 2 begins to choose action $R2$, the outcome $(R1, R2)$ will be achieved, and it is possible that fairness will be achieved by the end of period $t + 1$. Since the value of $\Delta U_+$ is positively proportional to the value of $\alpha_i$, we can see that the value of $\alpha_1$ can reflect the persistence of agent 1 to choose action $R1$ expecting for achieving fairness. The larger the $\alpha_1$ is, the more the number of time steps it will choose action $R1$. This conforms to the intuition that parameter $\alpha_i$ represents agent $i$'s suffering degree when its payoff is lower compared with other agents'. The larger the $\alpha_i$ is, the more willing it is to fight for fairness.

Suppose the outcome is $(R1, R2)$ at the end of the $t + 1$th period and agent 1's accumulated payoff has become higher than agent 2's. To achieve fairness, the outcome $(R2, R1)$ preferred by agent 2 has to be reached later. According to the adjustment strategy, agent 1 decreases its attitudes by $\Delta U_-$. Thus it becomes more risk-averse, which means that it will put more weights on the non-conflict ratio when choosing its action. When agent 2 tries to choose action $R1$ expecting for achieving outcome $(R2, R1)$, agent 1 will still choose action $R1$ for a number of time steps at first. If $\Delta U_-$ becomes larger, agent 1's attitudes will be decreased more, and agent 1 will stick to action $R1$ for lower number of time steps.[7] If $\Delta U_-$ becomes large enough and the number of time steps in that agent 1 chooses action $R1$ is lower than that of agent 2, agent 1 will begin to choose action $R2$, thus the outcome $(R2, R1)$ will be achieved, and fairness will be achieved later. Since the value of $\Delta U_-$ is positively proportional to the value of $\beta_i$, parameter $\beta_1$ can be understood as the reflection of the concession degree of agent 1 to give up its current action expecting for achieving fairness. This is also in accordance with the intuition in the fairness model that the parameter $\beta_i$ reflects agent $i$'s inequity-averse degree toward its advantage of payoff compared with other agents'. The larger the $\beta_i$ is, the less reluctant it is to give up its current action for achieving fairness.

### 3.1.4 Properties of the Adaptive Strategy

By enforcing our strategy, we are expecting that the outcomes in the repeated game alternate between the two pure strategy Nash equilibria $(R1, R2)$ and $(R2, R1)$, which results in fairness of the agents in the repeated game. However, the occurrence of fairness between the agents is dependent on the values of $\alpha_i$ and $\beta_i$, and we have

---

between $\text{attr}_{R1}$ and $\text{attr}_{R2}$ is changed to $\text{diff}' = \text{attr}_{R2} - \text{attr}_{R1} + \Delta U_+(h_{R1} - h_{R2}) + \Delta U_+(u_{R2} - u_{R1})$. Since $h_{R2} = 1$ at this time, $h_{R1} - h_{R2} \leq 0$, and also $u_{R2} - u_{R1} < 0$, we can easily see that $\text{diff}' < \text{diff}$.

[7] If agent 1 chooses its action depending on its non-conflict ratio only, the number of time steps that it sticks to action $R1$ is no larger than the length of history $h_l$.

the following theorems. Each theorem will also be verified by simulation in next section. Here $\alpha_i \geq 0$ and $0 \leq \beta_i < 1$ are assumed, which follows the ranges given in the fairness model and also allows the condition that $\alpha_i < \beta_i$ assuming that agent can suffer less from the inequity when its payoff is lower than the other agent's.

**Theorem 3.1** *If $\alpha_i$ and $\beta_i$ are both sufficiently large (for $i = 1, 2$), $\epsilon$-fairness between the agents will be achieved for any feasible value of $\epsilon$.*

*Proof* Without loss of generality, let us assume that the outcome is $(R1, R2)$ at the end of the period $t$, and agent 1's accumulated payoff and periodical payoff are both higher than agent 2's. To prove this theorem, we only have to show that the pure strategy Nash equilibrium $(R2, R1)$ is able to be achieved later, since agent 2 will receive higher payoff than agent 1 after the outcome $(R2, R1)$ is achieved and $\epsilon$-fairness will be achieved after certain time steps.

Let us analyze the range of the value of $\alpha_i$ first. According to our adjustment strategy, agent 2's attitude will be increased by $\Delta U_+$ to reduce the difference between its values of $\text{attr}_{R1}$ and $\text{attr}_{R2}$. If $\Delta U_+$ is too small such that $\text{attr}_{R1}$ is still smaller than $\text{attr}_{R2}$, agent 1 will not change its action, and the outcome will still be $(R1, R2)$. If $\Delta U_+$ is large enough such that $\text{attr}_{R1} > \text{attr}_{R2}$ is satisfied for agent 2, agent 2 will begin to choose action $R1$ instead and both agents will reduce their own attitudes on action $R1$ by $a_\Delta$ at each time step, since both agents choose $R1$ and result in conflict. The larger the value of $\Delta U_+$ becomes, the more increased agent 2's attitudes will be, and agent 2's value of $\text{attr}_{R1}$ will become much larger than its value of $\text{attr}_{R2}$, and thus agent 2 will stick to action $R1$ for more time steps. The maximum number of time steps $k_{\max}$ that agent 2 can stick to action $R1$ is obtained when its attitude on action $R1$ is maximized to 1. Let us assume that agent 1 sticks to action $R1$ for $k'$ times($k' < k_{\max}$) before agent 1 switches to action $R2$ when conflict occurs at the beginning of each period. The outcome $(R2, R1)$ will be able to be achieved as long as agent 2 can stick to action $R1$ for $k''$ times ($k'' > k'$) at the beginning of certain period. It is obvious that the sufficient condition to ensure it is that agent 2's increasing degree on attitude $a_{R1}$ is always larger than its decreasing degree on attitude $a_{R1}$, since agent 2's attitude on action $R1$ will be increased to its maximum value 1 eventually and it will be able to stick to action $R1$ for $k_{\max}$ times. Since the value of $\Delta U_+$ is positively proportional to the value of $\alpha_2$, we obtain the lower bound of $\alpha_2$ which ensures $\epsilon$-fairness can be achieved, when the following condition holds: $k_{\max} a_\Delta < \Delta U_+ = \alpha_2 \frac{(u_1 - u_2)}{u_2} (\alpha_2 > \frac{k_{\max} a_\Delta u_2}{u_1 - u_2})$, where $k_{\max} a_\Delta$ and $\Delta U_+$ are the maximum decreasing degree and the increasing degree on agent 2's attitude $a_{R1}$, respectively.

Similar to the above analysis, according to our adjustment strategy, agent 1 decreases its attitudes by $\Delta U_-$; thus the difference between its values of $\text{attr}_{R1}$ and $\text{attr}_{R2}$ will be decreased. If $\Delta U_-$ becomes larger, agent 1's attitudes will be decreased more, and the difference between agent 1's values of $\text{attr}_{R1}$ and $\text{attr}_{R2}$ will be decreased more; thus agent 1 will choose action $R1$ for lower number of time steps. Let us assume that agent 1 chooses action $R1$ for $k$ time steps at the beginning of the current period. From previous analysis, we know that the maximum number of time steps $k_{\max}$ that agent 2 can stick to action $R1$ is obtained when its attitude on action $R1$ is maximized to 1. To ensure that fairness can be achieved, the outcome $(R2, R1)$ must be able to be achieved. We can see that if agent 1's increasing degree on attitude $a_{R1}$ is smaller than its decreasing degree on attitude $a_{R1}$ within each period, its attitude on action $R1$ will be decreased continually, and the number of time steps $k$ it can stick to action $R1$ at the beginning of each period will be decreased continually as well. Therefore the outcome $(R2, R1)$ will be achieved when the value of $k$ is decreased to satisfy $k < k_{\max}$ and $\epsilon$-fairness can be achieved later. Since the value of $\Delta U_-$ is positively proportional to the value of $\beta_1$, we can obtain the lower bound of $\beta_1$ ensuring that $\epsilon$-fairness can be achieved, when the following condition holds: $(T - k_{\max})a_\Delta - k_{\max}a_\Delta < \Delta U_- = \beta_1 \frac{(u_1-u_2)}{u_1} (\beta_1 > \frac{((T-k_{\max})a_\Delta - k_{\max}a_\Delta)u_1}{u_1-u_2})$, where $T$ is the periodical length, $k_{\max}$ is the maximum number of times that agent 2 can stick to action $R1$ at the beginning of each period, and the equations on both sides are the minimum increasing degree and the decreasing degree on agent 1's attitudes, respectively.

If the outcome is $(R1, R2)$ at the end of the period $t$, we have to ensure that the outcome $(R2, R1)$ is able to be achieved later. We can get the lower bounds for the value of $\alpha_1$ and $\beta_2$ in a similar way and we omit it here. □

**Theorem 3.2** *If $\alpha_i$ is too small, no matter how large $\beta_i$ is (for $i = 1, 2$), $\epsilon$-fairness between the agents will not be achieved for any feasible value of $\epsilon$.*

*Proof* To prove $\epsilon$-fairness between the agents will not be achieved in this situation, we only need to show that only one of the two pure strategy Nash equilibria $(R1, R2)$ and $(R2, R1)$ will be achieved in the repeated game.

Let us assume that the outcome converges to $(R1, R2)$ at the end of period $t$, and agent 1 receives both higher accumulated payoff and periodical payoff than agent 2.

According to our adjustment strategy, agent 2's attitude will be increased by $\Delta U_+$; thus its difference between the values of $\text{attr}_{R1}$ and $\text{attr}_{R2}$ will be decreased ($\text{attr}_{R1} < \text{attr}_{R2}$ at this time). Besides, agent 1 decreases its attitudes by $\Delta U_-$; thus the difference between its values of $\text{attr}_{R1}$ and $\text{attr}_{R2}$ will be decreased ($\text{attr}_{R1} > \text{attr}_{R2}$ at this time). If $\alpha_2$ is too small, the value of $\Delta U_+$ will be too small. If $\text{attr}_{R1}$ is still smaller than $\text{attr}_{R2}$ after the adjustment, agent 2 will still choose action $R2$. Since $\beta_1 \in [0, 1)$ and $\frac{(u_1-u_2)}{u_1} \in [0, 1]$, we can have $\Delta U_- \in [0, 1)$. We also know that agent 1's attitude on action $R1$ is increased to 1 at the end of period $t$, and then its attitude on action $R1$ will be still larger than 0 after the adjustment, and thus we can see that agent 1 will still choose action $R1$ after adjustment even if $\beta_1$ has taken its maximum value. We can conclude that the outcome will still be $(R1, R2)$, and $\epsilon$-fairness will not be achieved either.

If the outcome is $(R2, R1)$ at the end of period $t$, we can get the conclusion that $\epsilon$-fairness cannot be achieved if $\alpha_1$ is too small no matter how large $\beta_2$ is in a similar way. □

**Theorem 3.3** *If $\beta_i$ is too small, no matter how large $\alpha_i$ is (for $i = 1, 2$), $\epsilon$-fairness between the agents will not be achieved for any feasible value of $\epsilon$.*

*Proof* Suppose the outcome converges to $(R1, R2)$ at the end of period $t$, and agent 2 receives both lower accumulated payoff and periodical payoff than agent 1. Based on our adjustment strategy, agent 1 decreases its attitudes by $\Delta U_{-}$, and thus the difference between its values of $\text{attr}_{R1}$ and $\text{attr}_{R2}$ will be decreased ($\text{attr}_{R1} > \text{attr}_{R2}$ at this time). The larger the $\Delta U_{-}$, the lower the number of time steps that agent 1 chooses action $R1$.

For agent 2, the maximum value of agent 2's attitude on action $R1$ is 1 no matter how large $\alpha_2$ is. From previous analysis, we know that the maximum number of time steps $k_{\max}$ that agent 2 can stick to action $R1$ is obtained when its attitude on action $R1$ is maximized to 1. Therefore we can see that the maximum number of time steps that agent 2 can stick to action $R1$ will be $k_{\max}$ no matter how large $\alpha_2$ is. If the value of $\beta_1$ is too small such that the number of time steps $k$ that agent 1 can stick to action $R1$ is larger than $k_{\max}$, then agent 2 will switch back to action $R2$ earlier than agent 1 and the outcome will still be $(R1, R2)$. This process will also repeat in the following periods infinitely since the decreasing of agent 1's attitude on action $R1$ at the beginning of each period will be compensated after no conflict occurs until the end of each period. Therefore we can get the conclusion that if $\beta_1$ is too small, $\epsilon$-fairness between the agents will not be achieved no matter how large $\alpha_2$ is.

If the outcome is $(R2, R1)$ at the end of period $t$ and agent 2 receives both higher accumulated payoff and periodical payoff than agent 1, we can get similar conclusion that if $\beta_2$ is too small, $\epsilon$-fairness between the agents will not be achieved no matter how large $\alpha_1$ is. □

**Theorem 3.4** *If the adaptive strategy without involving the inequity-averse fairness model is adopted for both agents, one of the pure strategy Nash equilibria will be converged to (eventually and always reached thereafter) after finite time steps in two-player conflicting-interest game.*

*Proof* Without loss of generality, let us take the two player's conflicting-interest game in Fig. 3.1 as an example to prove this theorem. According to the adaptive strategy, initially both of the agents choose action $R1$ since $\text{attr}_{R1} > \text{attr}_{R2}$ is satisfied, and conflict occurs; thus each agent's attitude and attractiveness for action $R1$ will be decreased. Assuming that agent 1 starts to choose $R2$ first ($\text{attr}_{R1} < \text{attr}_{R2}$), the outcome will become $(R2, R1)$. After this, no conflict occurs and thus agent 1's attitude $a^t_{R2}$ on action $R2$ and agent 2's attitude $a^t_{R1}$ on action $R1$ will both be increased at the end of each time step from now on. This guarantees that the outcome will stick to Nash equilibrium $(R2, R1)$ thereafter. Based on similar analyses, if agent 2 switches to action $R2$ first, the outcome will converge to Nash equilibrium $(R1, R2)$ thereafter. The last case is that both agents start to switch to action $R2$ simultaneously. From previous analysis, we have known that one of the pure strategy Nash equilibria will be converged to as long as it can be achieved occasionally. According to the adaptive strategy, we also know that both agents' attractiveness of these two actions will eventually be decreased to zero if both agents always choose the same action and alternate their action choice simultaneously, and also one of the pure strategy Nash equilibria can be reached due to random action selection of the agents. Therefore one of the pure strategy Nash equilibria can be definitely converged on in two-player conflicting-interest game, if the adaptive strategy without involving fairness model is adopted for both agents. □

## 3.1.5 Experimental Evaluations

### 3.1.5.1 Simulation

We perform a number of experiments to investigate the adaptive strategy and the influence of parameters $\alpha_i$ and $\beta_i$ on the performance of the strategy based on the fairness related criteria and verify that the theoretical analysis is consistent with the simulation results. The experiment setting follows the problem specification described in Sect. 3.1.2. The length of each period $T$ is preset to $T = 100$, and the game is repeated for 40 periods. We assume that both agents' non-conflict ratios on each action are initialized to 0 with history length $h_l = 3$. In our experiments, agent 2's attitudes are always set to be larger than agent 1's for the purpose of consistency to ensure that agent 2 always achieves higher payoff than agent 1 in the first period and each agent owns the same attitudes on both actions. The adjustment rates $a_\Delta$ for both agents are $a^1_\Delta = a^2_\Delta = 0.01$. Besides, $\alpha_1 = \alpha_2 = \alpha$ and $\beta_1 = \beta_2 = \beta$ are assumed here. The payoff matrix for the two agents' conflicting-interest game is instantiated as the same one shown in Fig. 3.1.[8]

We plot the outcome results of the repeated game and average accumulated payoff curves of both agents when $\alpha$ and $\beta$ are set to different values within their own ranges shown in Figs. 3.3, 3.4, 3.5, 3.6, 3.7, and 3.8. Figures 3.4 and 3.6 show

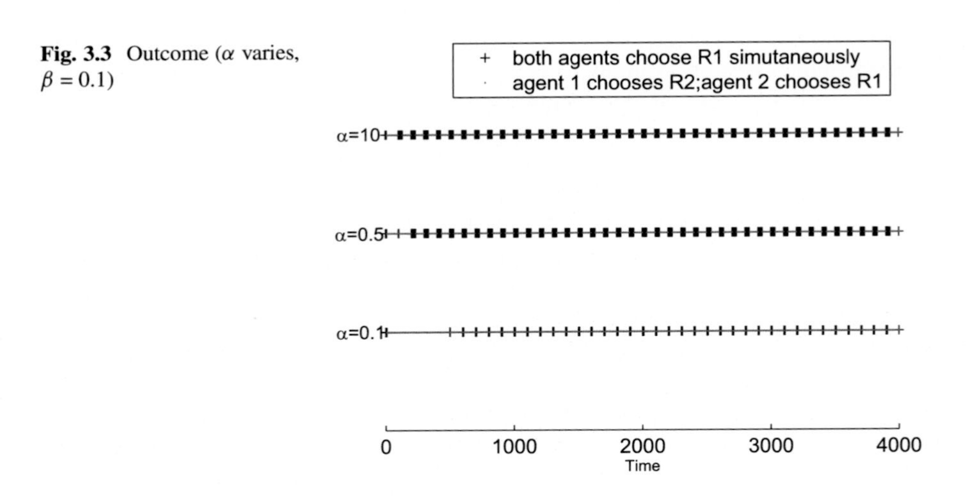


**Fig. 3.3** Outcome ($\alpha$ varies, $\beta = 0.1$)

[8]Since strict fairness ($\epsilon = 0$) can be achieved in this game, we will use the term fairness only in the following section.

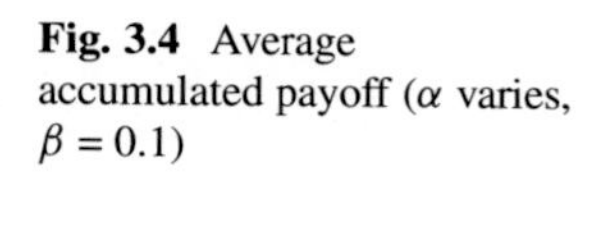

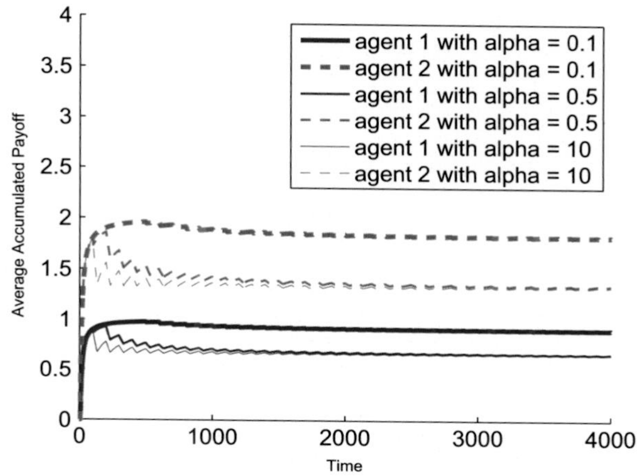


**Fig. 3.4** Average accumulated payoff ($\alpha$ varies, $\beta = 0.1$)

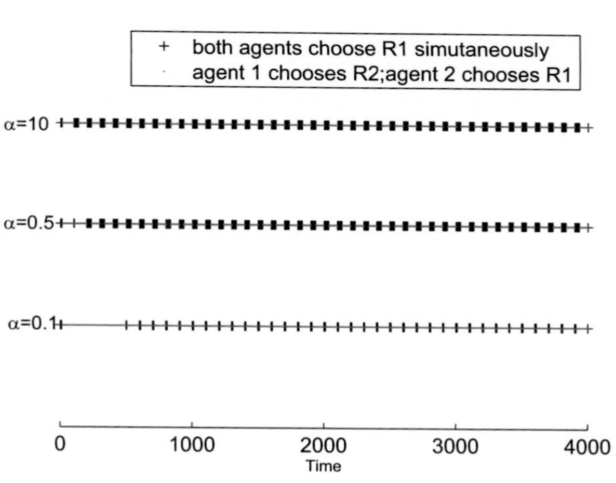


**Fig. 3.5** Outcome ($\alpha$ varies, $\beta = 0.5$)

the curves of each agent's average accumulated payoff when $\alpha$ is set to different values and $\beta$ is fixed to $\beta = 0.1$ and $\beta = 0.5$, respectively. In Figs. 3.4 and 3.6, agent 2 always obtains higher payoff than agent 1. This phenomenon can also be analyzed from the outcomes of the repeated game shown in Figs. 3.3, 3.4, and 3.5. In Figs. 3.3 and 3.5, we can see that agent 2 always chooses action $R1$ and agent 1 chooses action $R2$ most of the time, except there are a few time steps that agent 1 tries action $R1$ at the beginning of each period, which results in conflict, and then agent 1 returns to choose $R2$ after the unsuccessful trials for $R1$. Thus we can see that when the value of $\beta$ is small, fairness is never achieved in the repeated game no matter how large the value of $\alpha$ is.

Figure 3.8 shows the curves of each agent's average accumulated payoff when the value of $\alpha$ varies with $\beta$ fixed to its upper limit value $\beta = 1.0$. In this figure, we can see that when $\alpha$ is small (when $\alpha = 0.1$), agent 2 still always obtains higher

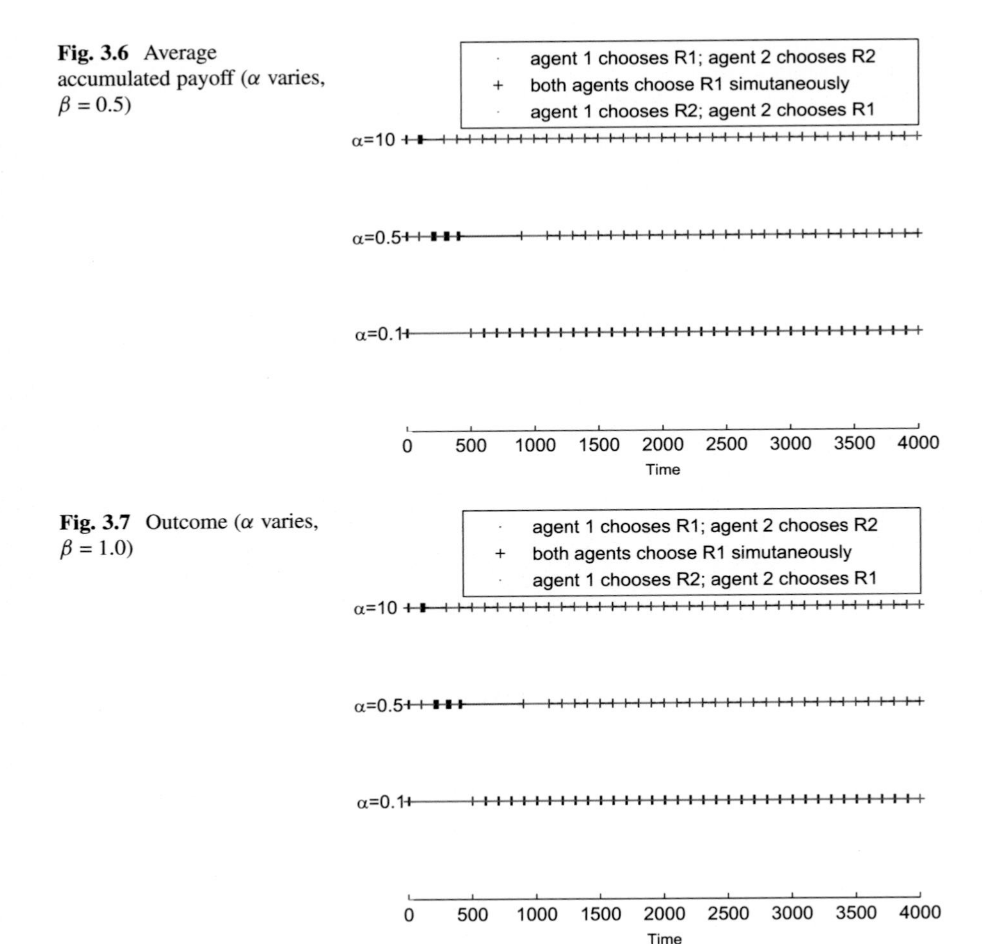


**Fig. 3.6** Average accumulated payoff ($\alpha$ varies, $\beta = 0.5$)

**Fig. 3.7** Outcome ($\alpha$ varies, $\beta = 1.0$)

payoff than agent 1. However, when $\alpha$ is larger ($\alpha = 0.5, 10$) shown in Fig. 3.8, agent 1 can obtain the same accumulated payoff as agent 2 after certain time steps. Therefore it implies that fairness can be achieved when the values of $\alpha$ and $\beta$ are both sufficiently large.

In Figs. 3.4, 3.5, 3.6, 3.7, and 3.8, we also can observe that agent 2 always achieves higher payoff than agent 1 when $\alpha = 0.1$ with the value of $\beta$ varying from 0.1 to 1. Thus fairness is not achieved between the agents in this situation. This verifies our theoretical analysis that fairness is never achieved when $\alpha$ is small no matter how large $\beta$ is.

When the values of $\alpha$ and $\beta$ are sufficiently large to ensure that fairness can be achieved between the agents, in Fig. 3.8, we can see that the value of time to fairness becomes smaller when the value of $\alpha$ becomes larger. This can be explained by the reason that the value of $\alpha$ reflects the agent's inequity-averse degree in an intuitive

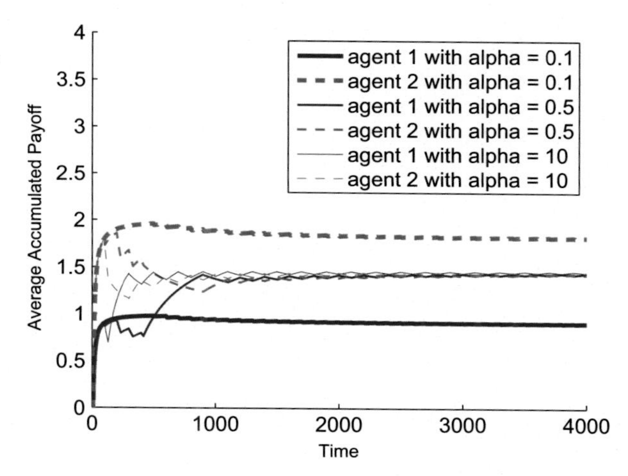


**Fig. 3.8** Average accumulated payoff ($\alpha$ varies, $\beta = 1.0$)

way: when the value of $\alpha$ is larger, agent 1 is more willing to stick to action $R1$ in order to achieve fairness; therefore fairness between the agents is achieved earlier.

Overall we have shown that the simulation results are in accordance with the theoretical properties from Theorems 3.1, 3.2, and 3.3 and thus verify the correctness of previous theoretical results.

#### 3.1.5.2 Comparison with Related Work

In this section, we compare the performance of the agents using the proposed adaptive strategy with the periodical policy in [8], based on the criteria proposed in Sect. 3.1.2.2.

##### Experimental Setting

The same conflicting-interest game is adopted, and we perform two sets of experiments with different period length $T$: (1) $T = 200$ (2) $T = 100$. For the adaptive strategy, the parameters $\alpha$ and $\beta$ for both agents are set to $\alpha = \beta = 1.0$ which are large enough to guarantee that fairness can be achieved. The settings of other parameters in the strategy are the same as those adopted in previous section's simulation. For the periodical policy of Nowé et al., the learning rate $\eta$ is $\eta = 0.1$, which is the same value adopted in [8]. Figures 3.9 and 3.10 show the average accumulated payoff each agent receives using the proposed adaptive strategy and the periodical policy when $T = 200$ and $T = 100$, respectively, and Figs. 3.11 and 3.12 show the corresponding outcome sequences of the repeated game when the proposed adaptive strategy and the periodical policy are adopted with $T = 200$ and $T = 100$, respectively.

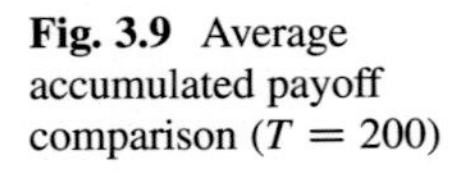

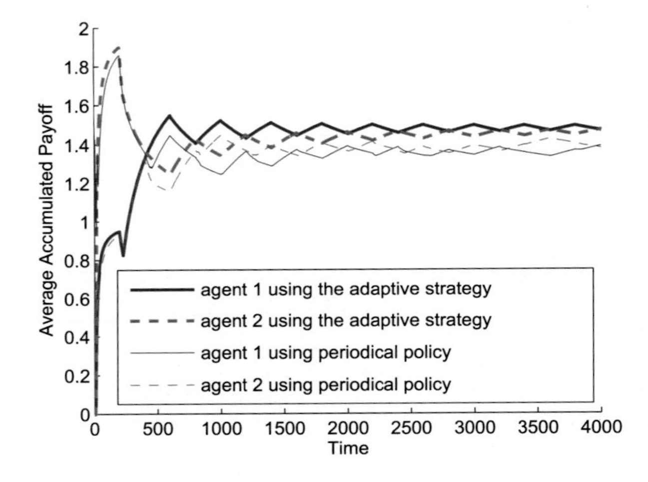


**Fig. 3.9** Average accumulated payoff comparison ($T = 200$)

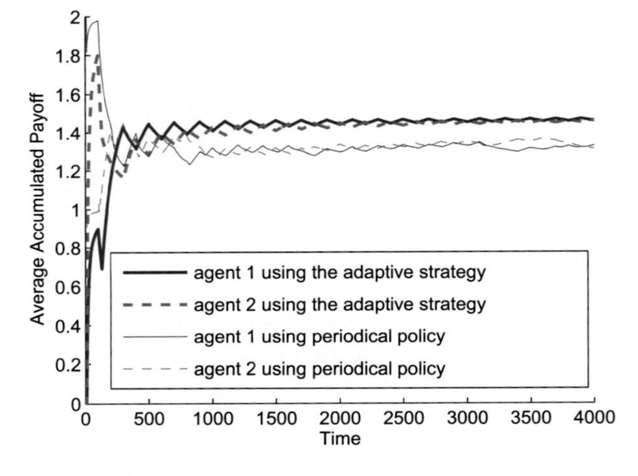


**Fig. 3.10** Average accumulated payoff comparison ($T = 100$)

### Time to Fairness

In Figs. 3.9 and 3.10, we can see that the agents can achieve fairness in the repeated game by adopting both strategies and the values of time to fairness are approximately the same for both of the strategies.

### Utilitarian Social Welfare and Length of Unfairness

In Figs. 3.9 and 3.10, we can observe that the average accumulated payoffs each agent receives by adopting the adaptive strategy are higher compared with those using periodical policy; thus utilitarian social welfare of the agents adopting the adaptive strategy is increased accordingly. This point can be illustrated further in

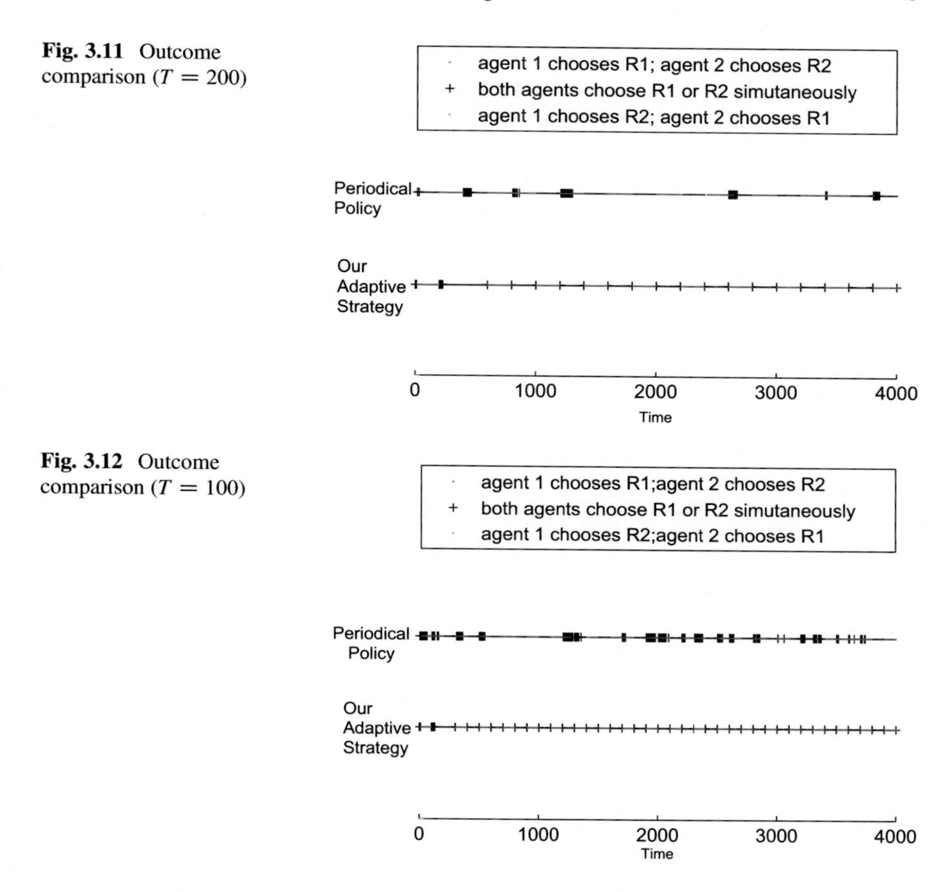


**Fig. 3.11** Outcome comparison ($T = 200$)

**Fig. 3.12** Outcome comparison ($T = 100$)

Figs. 3.11 and 3.12. The number of time steps with conflicting outcomes $(R1, R1)$ or $(R2, R2)$ is much smaller when the agents adopt the adaptive strategy than that when the periodical policy is adopted; thus less payoff is wasted in the conflicts. The rationality behind this is that in the adaptive strategy, the agents choose their actions in a more rational way without involving random exploration as the periodical policy and constantly exploit the best action adjusted by their individual risk attitudes; thus the chances of conflict occurring are greatly reduced.

Another observation is that the value of length of unfairness is approximately the same for both strategies when the game is repeated for certain time steps (starting from approximately 500 time steps), which is about twice of the length of the period $T$. When the length of unfairness becomes smaller by changing the value of period length $T$ from $T = 200$ to $T = 100$, each agent's average accumulated payoff adopting the periodical policy is decreased greatly. However, in the adaptive strategy, each agent's average accumulated payoff is only slightly reduced. The reason behind this is that for the periodical policy, the coordination of the agents is achieved by reinforcement learning, and the number of time steps with conflicting outcomes is higher than that in the adaptive strategy. Therefore the periodical policy is more sensitive to the change of the period length $T$. If $T$ is greatly decreased and the overall time of the repeated game is unchanged, the percentage of time steps that the agents achieve conflicting outcomes during reinforcement learning periods over the total time of the game will become much higher; thus this results in large decrease of the average accumulated payoffs for each agent. However, in the adaptive strategy, it only spends very few fixed number of time steps on conflicting outcomes at the beginning of each period, and the amount of payoff cost is not that sensitive to the change of the period length $T$; thus each agent's average accumulated payoff is only slightly decreased when $T$ is greatly reduced.

## 3.2 Game-Theoretic Fairness Models

In this section, we turn to consider approaching the goal of fairness from game-theoretic perspective. Game theory, a formal mathematical tool for game analysis, serves as a powerful framework to analyze the strategic interactions among agents. Classical game theory assumes that an individual agent is purely self-interested, which is commonly found to be inconsistent with the actual behaviors of human beings supported by numerous human experiments. Lots of evidences from human experiments suggest that *fairness* emotions play an important role in people's decision-making process, and thus it is important for us to explicitly incorporate fairness into game-theoretic frameworks. On one hand, to better explain and predict human behaviors, more refined game-theoretic models are needed by taking the effects of *fairness* motivation into account; on the other hand, game-theoretic models incorporating emotions like fairness in turn can provide us with valuable insights and instructions in designing more efficient and human aligned strategies for intelligence agents. To this end, in the book we make two major contributions as follows. First, in Sect. 3.2.1 we propose a fairness-based game-theoretic model for two-player normal-form games, which can better explain and predict actual human behaviors compared with previous fairness models. Second, we extend the descriptive fairness model (i.e., the inequity-averse fairness model) from single-shot games to infinitely repeated games with limit-of-means criterion in Sect. 3.2.2. We successfully show that there exist fairness equilibria under which the agents are guaranteed to achieve both efficient and fair outcomes.

### *3.2.1 Incorporating Fairness into Agent Interactions Modeled as Two-Player Normal-Form Games*

The typical assumption of individual agents in multiagent system is self-rationality, according to the principle of classical game theory. However, substantial evidences

**Fig. 3.13** Example 1: battle of the sexes[9]

| 1's payoff, 2's payoff | | Player 2's action | |
|---|---|---|---|
| | | Opera | Boxing |
| Player 1's action | Opera | 2X, X | 0, 0 |
| | Boxing | 0, 0 | X, 2X |

show that humans are not purely self-interested and thus this strong assumption of self-rationality has been recently relaxed in various ways [9, 14, 15]. One important aspect is that people strongly care about fairness. Since many multiagent systems are designed to interact with human or act on behalf of them, it is important that an agent's behavior should be (at least partially) aligned with human expectations, and failure to consider this may incur significant cost during the interactions. To assist the design of intelligent agents toward fairness, it is important for us to have some game-theoretic frameworks which can formally model fairness. Within these frameworks, we can formally analyze the strategic interactions among agents and predict agents' possible behaviors under the concern of fairness.

When people make decisions, apart from their material payoffs, they also care about the motive behind others' actions [16]. People are willing to sacrifice their own material interests to help those who are being kind and also punish those who are being unkind [17, 18]. Besides, people also care about the relative material payoffs with others and are willing to sacrifice their own material payoffs to move in the direction of decreasing the material payoff inequity [10]. Consider the battle-of-the-sexes game shown in Fig. 3.13, where $X > 0$. In this game, both agents prefer to take the same action together, but each prefers a different action choice. Specifically, agent 1 prefers that both agents choose action Opera together, while agent 2 prefers that both agents jointly take action Boxing.

In this game there exist two pure strategy Nash equilibria, i.e., (Opera, Opera) and (Boxing, Boxing). The outcomes (Opera, Boxing) and (Boxing, Opera) are not Nash equilibria, in which each agent always has the incentive to deviate from his current strategy to increase his material payoff. However, in reality, people usually do not make their decisions based on the material payoffs only. Apart from material payoff, people also take their fairness emotions into their decision-making processes. People not only care about their material payoff but also the way they are being treated by other people [17–19]. In reality, if agent 1 thinks that agent 2 is doing him a favor, then he will have the desire to do agent 2 a favor in return; if

[9]The payoffs specified in Examples 1 and 2 represent the agents' material payoffs.

agent 2 seems to be hurting agent 1 intentionally, agent 1 will be motivated to hurt agent 2 as well.

Based on the above theories, let us look at the outcome (Opera, Boxing) again. If agent 1 believes that agent 2 deliberately chooses Boxing instead of Opera which decreases his material payoff, he will have the incentive to hurt agent 2 in return in reality by sticking with Opera. If agent 2 is thinking in the same way, then he will also have the motivation to sacrifice his own interest and stick with Boxing to hurt agent 1. In this way, both agents are hurting each other to satisfy their emotional desires. If this kind of emotions is strong enough, then no agent is willing to deviate from the current strategy, and thus the outcome (Opera, Boxing) becomes an equilibrium.

Another important aspect of fairness emotions that can influence people's decisions is the fact that people not only care about their own material payoffs but also the relative material payoffs with others. It has been revealed that people show a weak aversion toward advantageous inequity and a strong aversion toward disadvantageous inequity [10, 20–22]. People may be willing to give up their material payoffs to move in the direction of more equitable outcomes. For example, consider the game shown in Fig. 3.14, where $X > 0$. It is easy to check that a Nash equilibrium is $(C, C)$. However, if we also consider the factor that people are inequity-averse, then agent 1 may be very unhappy about the current situation that his material payoff is much lower than that of agent 2. Thus agent 1 may have the incentive to switch to action $D$ to achieve an equitable outcome instead, and $(C, C)$ is not an equilibrium any more.

Furthermore, consider the outcome $(C, D)$, which is not a Nash equilibrium. If agent 1 believes that agent 2 intentionally chooses $D$ to avoid obtaining a higher material payoff than agent 1, agent 1 may wish to reciprocate by sticking with $C$, even though he could have chosen action $D$ to obtain a higher material payoff of $4X$ (if he chooses action $D$ and obtains $4X$, he may feel guilty for taking advantage of agent 2.). Similarly, if agent 2 also believes that agent 1 is doing a favor to him by choosing $C$ instead of $D$, he will also have the motivation to reciprocate by sticking with action $D$. Thus the outcome $(C, D)$ becomes an equilibrium if this kind of emotions between the agents is taken into consideration.

**Fig. 3.14** Example 2

| 1's payoff, 2's payoff | | Agent 2's action | |
|---|---|---|---|
| | | C | D |
| Agent 1's action | C | 4X, 10X | 2X, 2X |
| | D | 2X, 2X | 4X, X |

However, in previous fairness models [10, 16], only either of the previous factors is considered, which thus only reflects a partial view of the motivations behind fairness. To this end, we propose a fairness model within the game-theoretic framework, which combines both aspects of fairness motivations, to provide better predictions on the agents' behaviors.

#### 3.2.1.1 Fairness Model

Consider a two-player normal-form game, and let $A_1$, $A_2$ be the action spaces of agents 1 and 2. For each action profile $(a_1, b_2) \in A_1 \times A_2$, let $p_1(a_1, b_2)$ and $p_2(a_1, b_2)$ be the material payoff for agent 1 and 2, respectively.

Following the work of Fehr and Schmidt [10], to represent the inequity-averse degree of the agents, we assume that each agent $i$ has two inequity-averse factors $\alpha_i$ and $\beta_i$. These two factors represent agent $i$'s suffering degree when he receives relatively lower and higher material payoffs than his opponent, respectively. We use $\alpha_i'$ and $\beta_i'$ to denote agent $i$'s belief on agent $j$'s ($i \neq j$) inequity-averse factors and $\alpha_i''$ and $\beta_i''$ to denote his belief on agent $j$'s belief on agent $i$'s inequity-averse factors. The overall utility of agent $i$ thus depends on the following factors[10]: (1) the strategy profile played by the agents, (2) his inequity-averse factors, (3) his belief about agent $j$'s inequity factors, and (4) his belief about agent $j$'s belief about his own inequity-averse factors.

Each agent's inequity-averse factors and his higher-order beliefs model how he perceives the kindness from his opponent and also how kind he is to his opponent. Before formally defining these kindness functions, we adopt the inequity-averse model in [10] to define the *emotional utility* of agent $i$ as follows.

$$u_i(a_i, b_j) = p_i - \alpha_i \max\{p_j - p_i, 0\} - \beta_i \max\{p_i - p_j, 0\}, i \neq j \tag{3.4}$$

where $p_i$ and $p_j$ are the material payoffs of the agents $i$ and $j$ under strategy profile $(a_i, b_j)$ and $\alpha_i$ and $\beta_i$ are agent $i$'s inequity-averse factors.

Given that agent $j$ chooses strategy $b_j$, how kind is agent $i$ to agent $j$ if agent $i$ chooses strategy $a_i$? Based on Eq. 3.4, agent $i$ believes that agent $j$'s emotional utility is

$$u_j'(a_i, b_j) = p_j - \alpha_i' \max\{p_j - p_i, 0\} - \beta_i' \max\{p_i - p_j, 0\}. \tag{3.5}$$

Let $\prod(b_j)$ be the set of all possible emotional utility profiles if agent $j$ chooses action $b_j$ from agent $i$'s viewpoint, i.e., $\prod(b_j) = \{(u_i(a_i, b_j), u_j'(a_i, b_j)) \mid a_i \in A_i\}$. We define the reference emotional utility $u_j^e(b_j)$ as the expected value of agent $j$'s highest

[10]Note that it is possible to perform higher-level modeling here. However, as we will show, modeling at the second level is already enough and also can keep the model analysis tractable.

emotional utility $u_j^h(b_j)$ and his lowest one $u_j^l(b_j)$ among those Pareto-optimal points in $\prod(b_j)$. Besides, let $u_j^{\min}(b_j)$ be the minimum emotional utility of agent $j$ in $\prod(b_j)$.

Based on the above definitions, now we are ready to define the *kindness* of agent $i$ to agent $j$. This kindness function reflects how much agent $i$ believes that he is giving to agent $j$ with respect to the reference emotional utility.

**Definition 3.4** The *kindness* of agent $i$ to agent $j$ under strategy profile $(a_i, b_j)$ is given by

$$K_i(a_i, b_j, \alpha_i', \beta_i') = \frac{u_j'(a_i, b_j) - u_j^e(b_j)}{u_j^h(b_j) - u_j^{\min}(b_j)}. \tag{3.6}$$

If $u_j^h(b_j) - u_j^{\min}(b_j) = 0$, then $K_i(a_i, b_j, \alpha_i', \beta_i') = 0$.[11]

Similarly we can define how kind agent $i$ believes agent $j$ is to him. One major difference is that now agent $i$'s emotional utility is calculated based on his belief about agent $j$'s belief about his inequity-averse factors. Given a strategy profile $(a_i, b_j)$, agent $i$ believes that agent $j$ wants him to obtain emotional utility $u_i'(a_i, b_j)$, which is defined as follows:

$$u_i'(a_i, b_j) = p_i - \alpha_i'' \max\{p_j - p_i, 0\} - \beta_i'' \max\{p_i - p_j, 0\}. \tag{3.7}$$

Besides, we can similarly define $\prod(a_i) = \{(u_i'(a_i, b_j), u_j'(a_i, b_j)) \mid b_j \in A_j\}$, which is the set of all possible emotional utility profiles if agent $i$ chooses $a_i$ from agent $i$'s viewpoint. The reference utility $u_i^e(a_i)$ is the expected value of agent $i$'s highest emotional utility $u_i^h(a_i)$ and his lowest one $u_i^l(a_i)$ among those Pareto-optimal points in $\prod(a_i)$. Besides, let $u_i^{\min}(a_i)$ be the minimum emotional utility of agent $i$ in $\prod(a_i)$.

Finally, agent $i$'s *perceived kindness* from agent $j$ is defined as how much agent $i$ believes agent $j$ is giving to him.

**Definition 3.5** Player $i$'s *perceived kindness* from agent $j$ under strategy profile $(a_i, b_j)$ is given by

$$\delta_i(a_i, b_j, \alpha_i'', \beta_i'') = \frac{u_i'(a_i, b_j) - u_i^e(a_i)}{u_i^h(a_i) - u_i^{\min}(a_i)}. \tag{3.8}$$

If $u_i^h(a_i) - u_i^{\min}(a_i) = 0$, then $\delta_i(a_i, b_j, \alpha_i'', \beta_i'') = 0$.

Each agent's *overall utility* over an outcome $(a_i, b_j)$ is jointly determined by his emotional utility, his kindness, and perceived kindness from his opponent. Each agent $i$ always chooses action $a_i$ to maximize his own overall utility when he makes

---

[11] When $u_j^h(b_j) = u_j^{\min}(b_j)$, agent $j$ would always receive the same payoff no matter which action agent $i$ chooses, and thus there is no issue of kindness, and $K_i(a_i, b_j, \alpha_i', \beta_i') = 0$.

his decisions. Player $i$'s overall utility function is defined as follows:

$$U_i(a_i, b_j, \alpha_i, \beta_i, \alpha_i', \beta_i', \alpha_i'', \beta_i'') = u_i(a_i, b_j) + \\ K_i(a_i, b_j, \alpha_i', \beta_i') \times \delta_i(a_i, b_j, \alpha_i'', \beta_i''). \tag{3.9}$$

In this way, we have incorporated all the fairness motivations considered in previous discussions into the agents' decision-making processes. If agent $i$ believes that agent $j$ is kind to him, then $\delta_i(a_i, b_j, \alpha_i'', \beta_i'') > 0$. Thus agent $i$ will be emotionally motivated to show kindness to agent $j$ and choose an action $a_i$ such that $K_i(a_i, b_j, \alpha_i', \beta_i')$ is high to increase his overall utility. Similarly, if agent $i$ believes that agent $j$ is unkind to him ($\delta_i(a_i, b_j, \alpha_i'', \beta_i'') < 0$), then he will be emotionally motivated to show unkindness to agent $j$ and choose to minimize the value of $K_i(a_i, b_j, \alpha_i', \beta_i')$ to increase his overall utility. Besides, agents' kindness emotions reflect the agents' inequity-averse degrees. For example, assuming that the value of $\alpha_i$ is large, then agent $i$ will believe that agent $j$ is unkind to him if his material payoff is much lower than that of agent $j$ even if his current material payoff is already the highest among all his possible material payoffs. Thus agent $i$ will reciprocate by choosing an action $a_i'$ to decrease agent $j$'s emotional utility based on his own beliefs. Note that both kindness functions are normalized, and thus they are insensitive to the positive affine transformations of the material payoffs. However, the overall utility function is sensitive to such transformations. If the material payoffs are extremely large, the effect of $u_i(a_i, b_j)$ may dominate the agents' decisions and the influences of the agents' reciprocity emotions can be neglected, which is consistent with the evidences in human experiments [17, 23].

#### 3.2.1.2 Fairness Equilibrium

We define the concept of equilibrium using the concept of psychological Nash equilibrium defined by [24]. This is an analog of Nash equilibrium for psychological games with the additional requirement that all higher-order beliefs must match the actual ones.

**Definition 3.6** The strategy profile $(a_1, a_2) \in A_1 \times A_2$ is a *fairness equilibrium* if and only if the following conditions hold for $i = 1, 2, j \neq i$:

- $a_i \in \arg\max_{a_i \in A_i} U_i(a_i, a_j, \alpha_i, \beta_i, \alpha_i', \beta_i', \alpha_i'', \beta_i'')$
- $\alpha_i' = \alpha_j = \alpha_j''$
- $\beta_i' = \beta_j = \beta_j''$

In this fairness equilibrium definition, the first condition simply represents that each agent is maximizing his overall utility, while the second and the third conditions state that all higher-order beliefs are consistent with the actual ones. This solution concept is consistent with the discussions in previous examples. Considering Example 1, suppose that $\alpha_1 = \alpha_2 = \beta_1 = \beta_2 = 0.5$, and let the higher-order beliefs be consistent with these actual ones. For the outcome (Opera, Boxing), we have $K_1(\text{Opera}, \text{Boxing}, 0.5, 0.5) = -1$ and $\delta_i(\text{Opera}, \text{Boxing}, 0.5, 0.5) = -1$, and thus the overall utility of agent 1 is 1. If agent 1 chooses Boxing, then his overall utility will be $0.5X$. Thus if $X < 2$, agent 1 prefers Opera to Boxing given these beliefs. For the sake of symmetry, we can have that agent 2 would prefer Boxing to Opera in the same situation. For $X < 2$, therefore, (Opera, Boxing) is a fairness equilibrium. Intuitively, in this equilibrium, both agents are hostile toward each other and unwilling to concede to the other. If $X > 2$, the agents' desires for pursuing emotional utility will override their concerns for fairness; thus (Opera, Boxing) is not a fairness equilibrium. On the other hand, it is not difficult to check that both (Opera, Opera) and (Boxing, Boxing) are always equilibria. In these equilibria, both agents feel the kindness from the other and thus are willing to coordinate with the other for reciprocation, which also maximizes their emotional utilities.

Next let us consider Example 2, and the same set of values for the agents' inequity-averse factors is adopted as Example 1. For the outcome $(C, D)$, we have $K_1(C, D, 0.5, 0.5) = \frac{1}{2}$ and $\delta_i(C, D, 0.5, 0.5) = \frac{1}{2}$, and thus agent 1's overall utility is $2X + \frac{1}{4}$. If agent 1 switches to $D$, his kindness to agent $j$ is changed to $K_1(D, D, 0.5, 0.5) = -\frac{1}{2}$, and thus his overall utility becomes $2.5X - \frac{1}{4}$. Therefore, if $2X + \frac{1}{4} > 2.5X - \frac{1}{4}$, i.e., $X < 1$, he will prefer action $C$ to action $D$. Similarly, we can have the conclusion that if $X < 0.1$, agent 2 will prefer action $D$ to action $C$. Thus $(C, D)$ is a fairness equilibrium if $X < 0.1$. In this equilibrium, both agents are kind to the other by sacrificing their own personal interest. However, if the emotional utility becomes large enough such that the effect of the emotional utility dominates the overall utility, $(C, D)$ is not an equilibrium any more. For example, if $X > 1$, agent 1 will have the incentive to choose action $D$ to maximize his overall utility. Similarly, we can also check that $(C, C)$ is no longer an equilibrium any more for any $X > 0$, since agent 1's strong inequity-averse emotion always induces him to deviate to action $D$ to obtain an equal payoff.

Overall, we can see that in both examples, new equilibria can be introduced based on our fairness model, apart from the traditional Nash equilibria. Besides, some Nash equilibria are ruled out in certain situations (e.g., the Nash equilibrium $(C, C)$ is not a fairness equilibrium in Example 2). In next section, we are going to present some general conclusions about fairness equilibrium explaining why this is the case.

#### 3.2.1.3 General Theorems

For any game $G$, we can obtain its *emotional extension* $G'$ by replacing every material payoff profile with its corresponding emotional utility profile defined in Eq. 3.4. Considering the battle-of-the-sexes game, its emotional extension using the same set of parameters as in the previous section is shown in Fig. 3.15. If both agents play strategy Opera, then both of them are maximizing their opponents' emotional utilities simultaneously. Similarly, if agent 1 chooses strategy Opera, while agent 2 chooses Boxing, then both agents are minimizing their opponent's emotional

| 1's payoff, 2's payoff | | Agent 2's action | |
|---|---|---|---|
| | | Opera | Boxing |
| Agent 1's action | Opera | 1.5X, 0.5X | 0, 0 |
| | Boxing | 0, 0 | 0.5X, 1.5X |

**Fig. 3.15** Emotional extension of battle of the sexes

utilities at the same time. These strategy profiles are examples of *mutual-max* and *mutual-min* outcomes, which are formally defined as follows:

**Definition 3.7** A strategy profile $(a_1, a_2) \in A_1 \times A_2$ is a *mutual-max* outcome if, for $i \in \{1, 2\}, j \neq i, a_i \in \arg\max_{a \in A_i} u_j(a, a_j)$.

**Definition 3.8** A strategy profile $(a_1, a_2) \in A_1 \times A_2$ is a *mutual-min* outcome if, for $i = \{1, 2\}, j \neq i, a_i \in \arg\min_{a \in A_i} u_j(a, a_j)$.

For any game $G$, we can also define Nash equilibrium in its emotional extension $G'$ as follows:

**Definition 3.9** For any two-player normal-form game $G$, a strategy profile $(a_1, a_2) \in A_1 \times A_2$ is a Nash equilibrium in its emotional extension $G'$ if, for $i = 1, 2, j \neq i, a_i \in \arg\max_{a \in A_i} u_i(a, a_j)$.

We also characterize an outcome of a game based on the kindness function of the agents as follows:

**Definition 3.10** An outcome $(a_i, b_j)$ is:

- strictly (or weakly) positive if, for $i = 1, 2$, $K_i(a_i, b_j, \alpha'_i, \beta'_i) > 0$ (or $K_i(a_i, b_j, \alpha'_i, \beta'_i) \geq 0$);
- strictly (or weakly) negative if, for $i = 1, 2$, $K_i(a_i, b_j, \alpha'_i, \beta'_i) < 0$ (or $K_i(a_i, b_j, \alpha'_i, \beta'_i) \leq 0$);
- neutral if for $i = 1, 2$, $K_i(a_i, b_j, \alpha'_i, \beta'_i) = 0$;
- mixed if, for $i = 1, 2, i \neq j$, $K_i(a_i, b_j, \alpha'_i, \beta'_i) \times K_j(a_i, b_j, \alpha'_j, \beta'_j) < 0$.

Based on previous definitions, we are ready to state the sufficient conditions for a Nash equilibrium in game $G$'s emotional extension $G'$ to be a fairness equilibrium in the original game $G$.

**Theorem 3.5** *Suppose that the outcome $(a_1, b_2)$ is a Nash equilibrium in $G'$; if it is also either a mutual-max outcome or mutual-min outcome, then it is a fairness equilibrium in $G$.*

*Proof* Since $(a_1, b_2)$ is a Nash equilibrium in $G'$, both agents must maximize their emotional utilities in this outcome. If $(a_1, b_2)$ is also a mutual-max outcome, then both $K_1(a_1, b_2, \alpha'_1, \beta'_1)$ and $K_2(a_1, b_2, \alpha'_2, \beta'_2)$ must be nonnegative, which implies that both agents perceive kindness from the other. Therefore, for each agent, there is no incentive for him to deviate from the current strategy in terms of maximizing both his emotional utility and the satisfaction of his emotional reactions to the other agent (maximizing the product of his kindness factors). Thus the outcome $(a_1, b_2)$ is a fairness equilibrium. Similarly, if $(a_1, b_2)$ is a mutual-min outcome, then both $K_1(a_1, b_2, \alpha'_1, \beta'_1)$ and $K_2(a_1, b_2, \alpha'_2, \beta'_2)$ must be nonpositive. Therefore both agents have the incentive to stick with the current strategy to maximize the product of their kindness factors. Also their emotional utilities are maximized in $(a_1, b_2)$. Therefore their overall utilities are maximized which implies $(a_1, b_2)$ is a fairness equilibrium. □

Next we characterize the necessary conditions for an outcome to be a fairness equilibrium.

**Theorem 3.6** *If an outcome $(a_1, b_2)$ is a fairness equilibrium, then it must be either strictly positive or weakly negative.*

*Proof* This theorem can be proved by contradiction. Suppose that $(a_1, b_2)$ is a fairness equilibrium, and also we have $K_i(a_i, b_j, \alpha'_i, \beta'_i) > 0$, $K_j(a_i, b_j, \alpha'_j, \beta'_j) \leq 0$. Since $K_i(a_i, b_j, \alpha'_i, \beta'_i) > 0$, it indicates that agent $i$ is kind to agent $j$. Based on Definition 3.4, we know that there must exist another strategy $a'_i$ such that by choosing it, agent $i$ can increase his own emotional utility and also decrease agent $j$'s emotional utility at the same time. Besides, since $K_j(a_i, b_j, \alpha'_j, \beta'_j) \leq 0$, it implies that agent $j$ is treating agent $i$ in a bad or neutral way at least. Thus agent $i$ also emotionally has the incentive to choose strategy $a'_i$ rather than being nice to agent $j$. Overall, agent $i$ would have the incentive to deviate from strategy $a_i$ to maximize his overall utility. Thus $(a_1, b_2)$ is not an equilibrium, which contradicts with our initial assumption. The only outcomes consistent with the definition of fairness equilibrium, therefore, are those outcomes that are either strictly positive or weakly negative. □

Previous theorems state the sufficient and necessary conditions for an outcome to be a fairness equilibrium in the general case. Next we present several results which hold when the emotional utilities are either arbitrarily large or small. For the sake of convenient representation, given a game $G$, for each strategy profile $(a_1, b_2) \in A_1 \times A_2$, let us denote the corresponding pair of emotional utilities as $(X \times u_1^o(a_1, b_2), X \times u_2^o(a_1, b_2))$, where $X$ is a variable used as a scaling factor of the emotional utility. By setting the value of $X$ appropriately, therefore, we can get the corresponding game $G'(X)$ (or $G(X)$) of any scale.

Let us consider the emotional extension of the battle-of-the-sexes game in Fig. 3.15. Notice that (Opera, Boxing) is not a Nash equilibrium in the sense of the emotional utilities, but it is a strictly negative mutual-min outcome. If the value of $X$ is arbitrarily small ($X < 2$), the emotional utilities become unimportant, and the agents' emotional reactions begin to take control. As we have previously shown, in this condition, the outcome (Opera, Boxing) can be a fairness equilibrium. Similarly, we can check that, in Example 2, the outcome $(C, D)$ which is not a Nash equilibrium is strictly positive mutual-max. Besides, we have also shown that if the value of $X$ is small enough, $(C, D)$ becomes a fairness equilibrium. In general, we can have the following theorem.

**Theorem 3.7** *For any outcome $(a_1, b_2)$ that is either strictly positive mutual-max or strictly negative mutual-min, there always exists an $X'$ such that for all $X \in (0, X')$, the outcome $(a_1, b_2)$ is a fairness equilibrium in the game $G(X)$.*

*Proof* Suppose $(a_1, b_2)$ is strictly positive mutual-max, it implies that each agent $i$ is maximizing their kindness product $K_i(a_i, b_j, \alpha_i', \beta_i') \times \delta_i(a_i, b_j, \alpha_i'', \beta_i'')$. If this kindness factor dominates each agent's overall utility, then this outcome must be a fairness equilibrium, since their overall utilities are also maximized in this outcome. To achieve this, we only need to set the value of $X$ to be arbitrarily small such that the emotional utility becomes unimportant compared with the kindness product. Therefore, there always exists such a $X'$ such that for all $X \in (0, X')$, the outcome $(a_1, b_2)$ is a fairness equilibrium in game $G(X)$. Similarly we can prove for the case when $(a_1, b_2)$ is strictly negative mutual-min and it is omitted. □

Now let us consider another case when the value of $X$ is arbitrarily large. Consider the battle-of-the-sexes example again. As we have shown, if the value of $X$ is large enough, i.e., $X > 2$, then the emotional utility dominates, and thus the outcome (Opera, Boxing) is not a fairness equilibrium any more. This can be generalized to the following theorem.

**Theorem 3.8** *Given a game $G(X)$, if the outcome $(a_1, b_2)$ is not a Nash equilibrium in $G'(X)$, then there always exists an $X'$ such that for all $X > X'$, $(a_1, b_2)$ is not a fairness equilibrium in $G(X)$.*

*Proof* If the outcome $(a_1, b_2)$ is not a Nash equilibrium in $G'(X)$, then at least one agent has the incentive to deviate from his current strategy to increase his emotional utility. Besides, we know that the values of $K_i(a_i, b_j, \alpha_i', \beta_i')$ and $\delta_i(a_i, b_j, \alpha_i'', \beta_i'')$ are independent from the value of $X$. Thus it is obvious that when $X$ is arbitrarily large, this agent can increase his overall utility by deviating from his current strategy, and $(a_1, b_2)$ is not a fairness equilibrium any more. □

Similarly, we can have the symmetric theorem as follows.

**Theorem 3.9** *Given a game $G(X)$, if the outcome $(a_1, b_2)$ is a strict Nash equilibrium in $G'(X)$, then there always exists an $X'$ such that for all $X > X'$, $(a_1, b_2)$ is a fairness equilibrium in $G(X)$.*

For example, consider Example 2, it can be checked that the outcome $(D, C)$, which is a Nash equilibrium in Example 2's emotional extension, is always a fairness equilibrium when the value of $X$ is large enough. However, note that if an outcome is weak Nash equilibrium in $G'(X)$, there may not exist an $X$ such that it is a fairness equilibrium in $G(X)$.

Overall, we have theoretically analyzed and proved why some Nash equilibria can be ruled out and some non-Nash equilibria become fairness equilibria in different cases. Based on the above theorems, it can provide us with valuable instructions to identify fairness equilibria in games of different scales.

### 3.2.2 Incorporating Fairness into Infinitely Repeated Games with Conflicting Interests for Conflict Elimination

The underlying assumption of traditional game theory is individual rationality, and the solution concept of pure strategy Nash equilibrium has been commonly adopted for prescribing how the players should behave or predict how they will behave. In games without pure strategy Nash equilibrium, players may use mixed strategy or behavior strategy to make decisions in order to obtain higher expected payoffs. However, both of them have their own disadvantages in certain games. For example, consider a simple two-agent two-resource allocation problem which can be modeled as an anti-coordination game shown in Fig. 3.16a. Each action corresponds to the selection of one resource and each agent prefers using one resource on its own to the other one. In this game there exist two pure strategy Nash equilibria $(C, D)$ and $(D, C)$ and also one mixed strategy Nash equilibrium. If the players make their decisions according to either of the pure strategy Nash equilibria, it will always be unfair for one of the players since there is always one player receiving lower payoff than the other. The situation becomes even worse when the game is repeatedly played since their payoff difference is gradually enlarged. However, it can be particularly important to achieve the goal of fairness in these types of multiagent practical scenarios, since we are usually interested in maximizing the global performance of the system which is determined by the agent with the worst performance. If the players make their decisions by playing the mixed strategy Nash equilibrium, although they can obtain equal payoffs on average, it is extremely inefficient in terms of the payoffs they obtain due to the high probability of reaching conflicting outcomes $(C, C)$ or $(D, D)$.

| Player 1's payoff, Player 2's payoff | | Player 2's action | |
|---|---|---|---|
| | | C | D |
| Player 1's action | C | 0,0 | 9,1 |
| | D | 1,9 | 0,0 |

| 1's payoff, 2's payoff | | Player 2's action | |
|---|---|---|---|
| | | C | D |
| Player 1's action | C | 3, 3 | 0, 10 |
| | D | 10, 0 | 1, 1 |

**Fig. 3.16** Payoff matrices for anti-coordination game and a special case of prisoner's dilemma game. (**a**) Anti-coordination game. (**b**) A special case of prisoner's dilemma game

If the game is repeatedly played, it is not difficult to observe that the conflicts between achieving fairness and maximizing the players' payoffs can be resolved given that the outcomes $(C, D)$ and $(D, C)$ can be reached alternately. We can make use of trigger strategy to achieve this goal by making a prior agreement of achieving one round of $(C, D)$ and one round of $(D, C)$ alternately between the players and any player deviating from this agreement will be punished. It is not difficult to check this strategy profile is also in equilibrium under the assumption that each player is individually rational. However, one limitation is that it is not flexible and general since this specific trigger strategy is only applicable to this particular game. It fails to handle other kinds of games, e.g., when it is required to achieve one round of $(C, D)$ and two rounds of $(D, C)$ alternately to eliminate the conflicts between the players. More importantly, there is a lack of underlying intuition (i.e., why should the players punish others who deviate) behind trigger strategy.

To address the above issues, it is important to explicitly take into account the motive of fairness instead of replying on the assumption of individual rationality, which is not considered in traditional game theory. Researchers have found that fairness motive plays an important role in human decision-making process [17–19]. It is of significant value for us to incorporate fairness motive into the game-theoretic models, which can be used for guiding and predicting artificial agents' behaviors in order to get better alignment with actual human behaviors. Different game-theoretic models incorporating fairness [4, 10, 16, 25, 26] have been proposed to explain and predict human behaviors; however, these models are designed targeting at analyzing single-shot games such as ultimatum game and dictator game and thus cannot be directly applied to repeated games. However, multiagent scenarios usually involve repeated interactions among agents, which can be naturally modeled as repeated games. Accordingly it is equally important for us to take into account fairness in the context of repeated game from the game-theoretic perspective. To this end, we explicitly introduce the concept of fairness strategy and fairness strategy equilibrium in the context of infinitely repeated game by utilizing the inequity-averse fairness model [10] which is suitable for analyzing single-shot games only. Here we focus on repeated two-player games with conflicting interests showing how fairness strategy equilibrium can guide the agents to resolve the conflicts between them. Specifically, we illustrate that, using fairness strategy, the conflicts between achieving fairness and maximizing the players' payoffs in two-player conflicting-interest games can be eliminated. Besides, we prove that the pair of fairness strategies the players adopt is in equilibrium, and thus no player will have the incentive to unilaterally deviate.

#### 3.2.2.1 Analyzing Two-Agent Interaction Scenarios with Conflict Interests

In two-agent interaction scenarios with conflict interests, the most-preferred outcomes of the agents do not coincide. Traditional game theory offers us an elegant way of analyzing how individually rational agents will interact in such a situation; however, the analysis results are not satisfactory from the system's perspective. Here we focus on two-player two-action symmetric games with conflict interests, which can be classified into two different types: (1) each agent prefers one outcome on the top-right or bottom-left corner and both outcomes are pure strategy Nash equilibria; (2) each agent prefers one outcome on the top-right or bottom-left corner and either the outcome on the top-left or bottom-right corner is the only pure strategy Nash equilibrium. Next we point out the disadvantages of analyzing both types of games using the concepts of traditional game theory.

The first type of game can be represented by the anti-coordination (AC) game, and one concrete example is shown in Fig. 3.16a. In this game, there exist two pure strategy Nash equilibria in which both players choose the opposite actions. To some degree, it is reasonable for the players to play in one of the pure strategy Nash equilibria since the outcome will be stable and also the sum of these two players' payoffs is maximized. However, it is unfair for one of the players no matter which pure strategy Nash equilibrium is played, since in either case one player's payoff is always lower than the other player, and the situation becomes worse when the game is repeatedly played.

On the other hand, let us consider adopting mixed strategies.[12] In this game there also exists a mixed strategy Nash equilibrium wherein each player chooses actions $C$ and $D$ with probability 0.9 and 0.1, respectively. Since in essence a mixed strategy is probability measure over the action space, a player using mixed strategy needs to throw a dice to choose the action. If both players play in this mixed strategy Nash equilibrium, each player will obtain the payoff of 0 with a probability 0.82, the payoff of 9 with a probability 0.09, and the payoff of 1 with a probability of 0.09. Therefore we can see that it is highly possible (with probability of 0.82) that the players make unfortunate decisions such that they mis-coordinate on their action choices and they both obtain the lowest payoff of 0. Even though in this case the players can achieve equal payoffs 0.9 on average, it is extremely inefficient for the players to play in this mixed strategy Nash equilibrium due to the high mis-coordination. Furthermore, if the game is repeated for $t$ rounds, both players will get the payoff of $0.9t$ on average. Another weakness of adopting mixed strategy Nash equilibrium is that it fails to align with actual human behaviors, since in reality human do not throw a mental dice to choose their actions [27]. In contrast, a strategy which can better reflect people's actual decision-making process is highly desirable, since it enables an artificial agent's behaviors to better align with human behaviors during human-agent interactions.

The second type of game can be represented by the special case of prisoner's dilemma (PD) game, and we give an example shown in Fig. 3.16b. It is different from the commonly used PD game in that the temptation of deviation from mutual cooperation is much higher (e.g., payoff of 10) and thus mutual cooperation is not necessarily the best outcome to pursue when the game is repeatedly played. In contrast with the anti-coordination game, in this game there does not exist any mixed strategy Nash equilibrium. Same with the ordinary PD game, there only exists

[12]Note that we only discuss the case of mixed strategy here, since a mixed strategy is also a behavioral strategy in this simple example, and vice versa.

one pure strategy Nash equilibrium $(D, D)$, which yields a low payoff of 1 for each player. If the game is repeated for $t$ rounds, both players will only get the payoff of $t$ if playing in the pure strategy Nash equilibrium.

From previous analysis, we can see that, in conflicting-interest games, there exist some conflicts between the goal of achieving fairness[13] between players and the goal of maximizing the players' payoffs in analyzing results from traditional game theory. However, the conflicts can be resolved if the players can coordinate with each other to achieve their most-preferred outcomes alternately. Specifically, in the PD game, if the outcomes $(C, D)$ and $(D, C)$ are achieved alternately, then both players' payoffs are equalized, which are also the highest payoffs they can obtain without causing inequity between them. Similarly, in the same way the conflicts existing in the AC game can be eliminated as well. Therefore we are interested in the following question: does there exist a general strategy which can eliminate the conflicts by guaranteeing the players to achieve their most-preferred outcomes alternately and also retain the appealing property of stability, i.e., the pair of designed strategies is still in equilibrium? If so, the satisfying strategy can provide valuable instructions for agent designers: we can simply design our agents to behave in the way specified by this satisfying strategy and the system is both efficient and stable which is guaranteed by the property of this strategy. To this end, we propose a novel way for players to make their decisions by incorporating the motive of fairness observed from humans, which will be introduced in next section.

#### 3.2.2.2 Fairness Strategy and Fairness Strategy Equilibrium in Infinitely Repeated Games

In this section, we introduce the concepts of fairness strategy and fairness strategy equilibrium in the context of infinitely repeated games. A strategic game is a game where each player is asked to choose an action simultaneously and the game is played once and for all. If a strategic game is infinitely repeated, it is called an infinitely repeated game. Traditional game theory is based on the assumption of individual rationality, that is, it is assumed that the players are exclusively pursuing their own material payoffs. However, substantial evidence suggests that fairness motive plays an important role in human's decision-making process [17–19]. People usually prefer to those fair outcomes in which their payoffs are more equalized instead of those outcomes maximizing their individual payoff only. One of the most influential models for explaining the fairness behaviors of human is the inequity-averse theory [10]. According to this theory, in human decision-making process, people usually make their decisions not only based on material payoffs but also the payoff difference with their opponents. When a person receives different (higher

[13]Here we simply say that fairness is achieved if the players' payoffs are equal since we focus on symmetric games only. More general definition of $\epsilon$-fairness [2] may be adopted in general-sum games as future work.

or lower) payoff from others, the perceived utility of this person can be different from the original material payoff he receives. Under this model, people are still assumed to be rational and they pursue the maximization of their perceived utilities instead of their material payoffs. Based on the inequity-averse theory [10], in a two-player interaction, player $i$'s perceived utility $U_i(a_i, a_j)$ for action profile $(a_i, a_j)$ is calculated as follows:

$$U_i(a_i, a_j) = u_i - \alpha_i \max\{u_j - u_i, 0\} - \beta_i \max\{u_i - u_j, 0\}, i \neq j \quad (3.10)$$

where $u_i$ and $u_j$ are the corresponding material payoffs of players $i$ and $j$ under the action profile $(a_i, a_j)$, the second term describes the utility cost when he receives lower payoff than player $j$, and the third term measures the utility cost when he receives higher payoff than player $j$. The two parameters $\alpha_i$ and $\beta_i$ are the corresponding weighting factors for the last two terms and represent player $i$'s suffering degrees when it receives lower and higher payoff than player $j$, respectively.

In Eq. 3.10, we can see that the values of $\alpha_i$ and $\beta_i$ can have significant effects on player $i$'s decision-making process. For example, consider the strategy profile $(C, D)$ in the variant of prisoner's dilemma game in Fig. 3.16b, in which player 2 receives much higher material payoff than player 1. If the value of $\beta_2$ is large enough such that player 2's perceived utility $U_2$ in the outcome $(C, D)$ becomes lower than 3, then player 2 may prefer to choose action $C$ to achieve the outcome $(C, C)$ to increase his perceived utility. For different players, they may exhibit different degrees of emotional reactions toward the same amount of unequal payoffs and accordingly each player $i$'s values of $\alpha_i$ and $\beta_i$ can be quite different. We can view $\alpha_i$ and $\beta_i$ as the instinctive property of player $i$ which reflects his fairness motive degree. To formally capture this, we associate a *fairness attitude* to each player $i$, which is represented by the pair of values $(\alpha_i, \beta_i)$.

**Definition 3.11** The *fairness attitude* $f_i^t$ of a player $i$ at time $t$ is a pair of values $(\alpha_i^t, \beta_i^t)$, where $\alpha_i^t$ and $\beta_i^t$ are the weighting factors for the cases when the player's payoff is lower and higher than others, respectively. Here $\alpha_i^t$ and $\beta_i^t$ are real numbers in $(-\infty, +\infty)$.[14]

Following the theory of inequity-aversion [10], we suggest that at each round $t$, each player makes his decision based on his fairness attitude $f_i^t$. Specifically, each player calculates his attractiveness of each action profile based on his own fairness attitude according to Eq. 3.11, which is a natural extension of Eq. 3.10 from one-shot games to the context of repeated games.

$$\text{Attr}_i^t(a_i, a_j) = cu_i^t - \alpha_i^t \max\{cu_j^t - cu_i^t, 0\} - \beta_i^t \max\{cu_i^t - cu_j^t, 0\}, i \neq j \quad (3.11)$$

[14]Note that here the ranges of $\alpha_i^t$ and $\beta_i^t$ cover the negative ranges since there exist some people who prefer to see other people suffer (have less payoff than himself) or be altruistic (give other people more payoff).

where $cu_i^t$ and $cu_j^t$ are the accumulated payoff of players $i$ and $j$ by round $t$ if the outcome $(a_i, a_j)$ is achieved in this round. In this way, the original game matrix can be transformed into an attractiveness matrix wherein the values for each entry corresponding to action profile $(a_i, a_j)$ are changed to $(\text{Attr}_i^t(a_i, a_j), \text{Attr}_j^t(a_i, a_j))$. Based on this attractiveness matrix, at each round $t$ each player will choose the action with the highest attractiveness in response to the action chosen by the other player.

During the repeated interaction with others, people have a desire to reward a good intention (or punish a bad intention) by acting as if they dislike being better off (or worse off) [28]. This implies that the values of $\alpha_i$ and $\beta_i$ in player $i$'s fairness attitude can be changed dynamically [10]. Since each player makes decision based on his fairness attitude at each round, his behaviors can become quite different as his fairness attitude changes. To model this in the context of repeated games, we define the *fairness strategy* of player $i$ to be a function that assigns values to $(\alpha_i, \beta_i)$ in his fairness attitude $f_i$ at the end of each round $t$.

**Definition 3.12** A *fairness strategy* $FS_i$ of a player $i$ in an infinitely repeated game is a function which assigns a fairness attitude $f_i^t$ to this player at the end of each round $t$ of the infinitely repeated game.

Suppose that player $i$ and player $j$ employ fairness strategies $FS_i$ and $FS_j$, respectively, if both players' current fairness strategies are in the best response of his opponent's, we call this *fairness strategy equilibrium* (FSE). In other words, when the pair of fairness strategies the players are adopting is in fairness strategy equilibrium, neither of them will have the incentive to unilaterally deviate from it.

**Definition 3.13** A *fairness strategy equilibrium* of an infinitely repeated $2 \times 2$ symmetric game is a fairness strategy profile $FS^*$ satisfying the property that for players $i, j \in N, i \neq j$, we have $O(FS_i^*, FS_j^*) \succsim_i O(FS_i', FS_j^*)$ for any fairness strategy $FS_i$ of player $i$, where $O(FS_i^*, FS_j^*)$ and $O(FS_i', FS_j^*)$ are the outcomes of the infinitely repeated game under the pair of fairness strategy $(FS_i^*, FS_j^*)$ and $(FS_i', FS_j^*)$, respectively, and the meaning of $\succsim_i$ follows the limit of means preference relation.

#### 3.2.2.3 Fairness Strategy and Fairness Strategy Equilibrium in Infinitely Repeated Games with Conflicting Interests

In this section, we show how to apply fairness strategy and fairness strategy equilibrium to infinitely repeated games with conflicting interests to guide and analyze the players' decision-making processes. In contrast with pure strategy or mixed strategy Nash equilibria, we will show that there exists a general pair of fairness strategies in equilibrium (FSE) under which the two players' most-preferred outcomes can be achieved alternately. In this way, the conflict between achieving fairness and maximizing each player's payoff can be resolved. Besides, no player has the incentive to unilaterally deviate from it since it is in equilibrium. The general

**Fig. 3.17** General form of AC game and PG game

| 1's payoff, 2's payoff | | Player 2's action | |
|---|---|---|---|
| | | C | D |
| Player 1's action | C | a, a | c, d |
| | D | d, c | b, b |

form of two-player two-action symmetric game with conflicting interests can be represented in Fig. 3.17 with the constraints of $c > d > a, c > b, c+d > 2a, c+d > 2b$ and $c > a > b > d$, and $c+d > 2a$, respectively.[15] Without loss of generality, we assume that the values of all entries are positive. For ease of exposition, we assume that the initial payoffs of both agents are $cu_1^0 = cu_2^0 = 0$.

We construct a pair of fairness strategies shown in Eqs. 3.12 and 3.13, respectively.

$$
f_i^t = \begin{cases} \alpha_i^t = \frac{d-a}{c-d} + 0.1, \beta_i^t = \frac{c-b}{c-d} - 0.1 & \text{if } t = 1 \\ \alpha_i^t = \alpha_i^{t-1}, \beta_i^t = \beta_i^{t-1} + \frac{c+d-a-b}{c-d} + 0.2 & \text{if } t > 1 \text{ and } cu_i^{t-1} > cu_j^{t-1} \text{ and } cu_i^{t-2} \le cu_j^{t-2} \\ \alpha_i^t = \alpha_i^{t-1}, \beta_i^t = \beta_i^{t-1} & \text{if } t > 1 \text{ and } cu_i^{t-1} \le cu_j^{t-1} \text{ and } cu_i^{t-2} \le cu_j^{t-2} \\ \alpha_i^t = \alpha_i^{t-1}, \beta_i^t = \beta_i^{t-1} & \text{if } t > 1 \text{ and } cu_i^{t-1} > cu_j^{t-1} \text{ and } cu_i^{t-2} > cu_j^{t-2} \\ \alpha_i^t = \alpha_i^{t-1}, \beta_i^t = \beta_i^{t-1} - \frac{c+d-a-b}{c-d} - 0.2 & \text{if } t > 1 \text{ and } cu_i^{t-1} \le cu_j^{t-1} \text{ and } cu_i^{t-2} > cu_j^{t-2} \end{cases} \tag{3.12}
$$

$$
f_j^t = \begin{cases} \alpha_j^t = \frac{d-a}{c-d} - 0.1, \beta_j^t = \frac{c-b}{c-d} + 0.1 & \text{if } t = 1 \\ \alpha_j^t = \alpha_j^{t-1} - \frac{c+d-a-b}{c-d} - 0.2, \beta_i^t = \beta_j^{t-1} & \text{if } t > 1 \text{ and } cu_j^{t-1} \ge cu_i^{t-1} \text{ and } cu_j^{t-2} < cu_i^{t-2} \\ \alpha_j^t = \alpha_j^{t-1}, \beta_j^t = \beta_j^{t-1} & \text{if } t > 1 \text{ and } cu_j^{t-1} < cu_i^{t-1} \text{ and } cu_j^{t-2} < cu_i^{t-2} \\ \alpha_j^t = \alpha_j^{t-1}, \beta_j^t = \beta_j^{t-1} & \text{if } t > 1 \text{ and } cu_j^{t-1} \ge cu_i^{t-1} \text{ and } cu_j^{t-2} \ge cu_i^{t-2} \\ \alpha_j^t = \alpha_j^{t-1} + \frac{c+d-a-b}{c-d} + 0.2, \beta_j^t = \beta_j^{t-1} & \text{if } t > 1 \text{ and } cu_j^{t-1} < cu_i^{t-1} \text{ and } cu_j^{t-2} \ge cu_j^{t-2} \end{cases} \tag{3.13}
$$

This pair of fairness strategies can be understood intuitively based on how the players' fairness attitudes change with round as follows. Player $i$ initially acts as a bad player such that he feels little guilty when he obtains higher payoff than his opponent player and also becomes extremely annoyed when his payoff becomes lower than his opponent. However, if his payoff becomes higher than his opponent and thus he is aware of the kind intention from his opponent, he will be converted to a good player whose behavior is in opposite with that of a bad player. Otherwise

[15] Note that the notation here is a little different from the example PD game in Fig. 3.16b, since under this general constraints, the Nash equilibrium in the PD game is $(C, C)$ instead of $(D, D)$.

he will become a bad player again. On the contrary, player $j$ initially behaves as a good player. He will change to a bad player if he receives lower cumulated payoff than his opponent (i.e., getting the impression that his opponent is unkind) and also change back to a good player otherwise.

Next we are going to show that if the agents make their decision following this pair of fairness strategies, then the conflicts between achieving fairness and maximizing the sum of both player's payoffs can be resolved and also prove that it is indeed a fairness strategy equilibrium in two cases.

### Infinitely Repeated AC Game

Supposing the players are playing the infinitely repeated AC game using the pair of fairness strategies given in Eqs. 3.12 and 3.13. In the first round, based on Eq. 3.11 the original game matrix can be transformed to the attractiveness matrix shown in Fig. 3.18. Suppose player 2 chooses action $C$; it is more attractive for player 1 to choose $C$ since $a > a - 0.1c + 0.1d$ (based on the relation of $c > d$); if player 2 chooses action $D$, the best response for player 1 is also $C$ since $b + 0.1c - 0.1d > b$. Thus player 1 will choose action $C$ in the first round. Similarly, we can get the conclusion that player 2 will choose action $D$ in the first round, and the outcome in round 1 is $(C, D)$.

The accumulated payoffs of both players after round 1 are $cu_1^1 = c, cu_2^1 = d$, respectively. Accordingly, at the beginning of round 2, both players will adjust their fairness attitudes according to their fairness strategies. The transformed

| Player 1's payoff, Player 2's payoff | | Player 2's action | |
|---|---|---|---|
| | | C | D |
| Player 1's action | C | a, a | b+0.1c-0.1d, a+0.1c-0.1d |
| | D | a-0.1c+0.1d, b-0.1c+0.1d | b, b |

**Fig. 3.18** The transformed game matrix of AC game in round 1

| Player 1's payoff, Player 2's payoff | | Player 2's action | |
|---|---|---|---|
| | | C | D |
| Player 1's action | C | -1.1c-0.9d+2a+2b, -1.1c-0.9d+3a+b | -2.2c-1.8d+2a+2b, -2.2c-1.8d+4a+2b |
| | D | c+d, c+d | -1.1c-0.9d+a+3b, -1.1c-0.9d+2a+2b |

**Fig. 3.19** The transformed game matrix of AC game in round 2

attractiveness matrix based on the new pair of fairness attitudes is shown in Fig. 3.19. Suppose player 2 chooses $C$, based on simple calculations, we can check that $-1.1c-0.9d+2a+2b < c+d$, and thus the best response for player 1 is $D$. If player 2 chooses $D$, it can be verified that $-2.2c-1.8d+2a+2b < -1.1c-0.9d+a+3b$ and thus it is still in player 1's best interest to choose action $D$. Therefore player 1 chooses action $D$ in round 2. Based on similar analysis, we can know that player 2 chooses action $C$. Therefore the outcome at the end of round 2 is $(D, C)$. At the end of round 2, both players' accumulated payoffs become equal, that is, $cu_1^2 = cu_2^2 = c + d$. It is not difficult to see that the transformed game matrix in round 3 is similar with round 1 except that the value of each entry is increased by a constant $c + d$, and thus the analysis result will be the same as round 1. Similar analysis can be repeated for the following rounds as both agents keep adjusting their fairness attitudes accordingly, and the outcomes of this repeated game will be $(C, D), (D, C), (C, D), (D, C), \ldots$, and so on. The total payoffs of each player after $t$ rounds of the game ($t$ is even) are $(c + d)t/2$. Note that this is the maximum equal payoff that the players can obtain in this infinitely repeated game. For comparison, the players can only obtain extremely unequal payoff of $ct$ and $dt$, respectively, under either of the pure strategy Nash equilibria and much lower equal payoff of $\frac{cd-ab}{c+d-a-b}t$ on average under the mixed strategy Nash equilibrium.

**Theorem 3.10** *For any two-player anti-coordination game G represented in Fig. 3.17 satisfying the constraints of* $c > d > a, c > b, c + d > 2a, c + d > 2b$, *there always exists a fairness strategy equilibrium in the infinitely repeated version of G which can resolve the conflicts between achieving fairness and maximizing the sum of the players' payoffs by guaranteeing both players' most-preferred outcomes are achieved alternately.*

*Proof* We only need to show that the fairness strategy profile in Eqs. 3.12 and 3.13 is a fairness strategy equilibrium in the infinitely repeated AC game. Without loss of generality, consider any odd round $t$; we have known that the outcome when the players are using this pair of fairness strategies is $(C, D)$. Since $(C, D)$ is one pure strategy Nash equilibrium in this game, no agent will have the incentive to unilaterally deviate from this fairness strategy to choose another action in round $t$. Otherwise, a lower payoff will be obtained due to mis-coordination. We can get the same conclusion for any even round $t$ as well. Thus the pair of fairness strategies in Eqs. 3.12 and 3.13 is a FSE in this infinitely repeated AC game. □

### Infinitely Repeated PD Game

First let us consider how the players will make decisions in the infinitely repeated PD game using the pair of fairness strategies in Eqs. 3.12 and 3.13. Since the one-shot PD game shares its general form with AC game and also we are considering the same pair of fairness strategies, it is not difficult to see that the transformed game matrix in the first round is the same with that in AC game shown in Fig. 3.18. Based on the relation constraints of $c > a > b > d, c + d > 2a$, we can easily know that the outcome in round 1 is $(C, D)$ following the same analysis in Sect. 3.2.2.3.

At the beginning of round 2, the accumulated payoffs of both players are $cu_1^1 = c, cu_2^1 = d$. Again, since we are considering the same fairness strategy as the one adopted in AC game and both games are represented by the same general form, the transformed game matrix in round 2 is the same with the one in AC game shown in Fig. 3.19. According to the relation constraints $c > a > b > d, c + d > 2a$, we can come to the conclusion that player 1 will choose action $D$ and player 2 will choose action $C$; and thus the outcome in round 2 is $(D, C)$.

As both players adjust their fairness attitude and their choice of actions accordingly in each round, the outcome of this repeated game will be $(C, D), (D, C), (C, D), (D, C), \ldots$, and so on. The total payoffs of each agent after $t$ rounds of the game ($t$ is even) are $\frac{c+d}{2}t$. Besides, we can see that this is the maximum payoff that both players can obtain without causing any inequity between them.

**Theorem 3.11** *For any two-player variance of prisoner's dilemma game G represented in Fig. 3.17 satisfying the constraints of* $c > a > b > d, c + d > 2a$, *there always exists a fairness strategy equilibrium in the infinitely repeated version of G which can resolve the conflicts between achieving fairness and maximizing the players' payoffs by guaranteeing both players' most-preferred outcomes are achieved alternately.*

*Proof* We need to prove that the pair of fairness strategies in Eqs. 3.12 and 3.13 is in fairness strategy equilibrium in infinitely repeated PD game. We know that the outcome sequence under this pair of fairness strategies is $(C, D), (D, C), (C, D), (D, C), \ldots$, and so on. However, different from AC game, neither the outcome $(C, D)$ nor $(D, C)$ is a Nash equilibrium; it is possible that the players may deviate from them to increase their one-round payoffs in different ways. Thus we

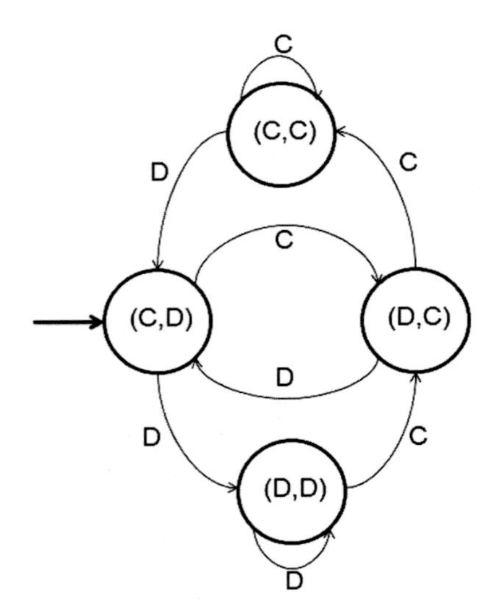


**Fig. 3.20** The finite state automaton of the infinitely repeated PD game

need to prove that both players cannot increase their individual payoffs no matter how to deviate. To make the proof clear, we represent the players' every possible dynamic decision-making path in this game using a finite state automaton shown in Fig. 3.20. Without loss of generality, this finite state automation is constructed from player 2's viewpoint. Each circle is a state in the automaton and its label is the outcome in that state. The transactions are represented as labeled arrows, and the label of each arrow is the action player 2 will choose. The unlabeled arrow indicates entering the initial state. As analyzed earlier, if both players are adopting this pair of fairness strategy, the outcomes will alternate between $(C, D)$ and $(D, C)$ which is represented by the middle loop between the two states $(C, D)$ and $(D, C)$. If player 2 deviates his fairness strategy, there are two ending states where he starts deviating, that is, state $(C, D)$ or state $(D, C)$.

If player 2 starts deviating from his fairness strategy when leaving state $(D, C)$, there are three possible situations to end with. The first one is to switch to choose $C$ and stick to it forever; thus the outcomes thereafter will always be $(C, C)$, which is indicated by being stuck in the most-upper self-loop on state $(C, C)$. The second situation is that player 2 deviates from his fairness strategy by choosing action $C$ (transiting from state $(D, C)$ to state $(C, C)$) and chooses action $D$ (transiting to state $(C, D)$) after staying in state $(C, C)$ for finite rounds, and finally he adopts his fairness strategy again forever (i.e., ending in the middle loop between state $(C, D)$ and $(D, C)$). The final situation is that player 2 always deviates and returns to his fairness strategy endlessly in an alternate way, which can be represented by the upper half loop $(D, C) \rightarrow (C, C) \rightarrow (C, D) \rightarrow (D, C)$ in the automaton. Next we will formally show that all the three deviating situations are no better than the non-deviating situation for player 2 separately.

Suppose that in round $k$ player 1 chooses $C$ according to the fairness strategy in Eq. 3.12 and player 2 unilaterally deviates from his fairness strategy and chooses $C$ in order to increase his one-round payoff (i.e., obtain payoff $a$ instead of $d$ ($a > d$)). For the first deviating situation, player 2 can get the total payoff of $(k/2) * (c + d) + a + (t - k)a = \frac{(c+d)k}{2} + (t - k + 1)a$ in t rounds of the game. As $\lim_{t\to\infty}(\frac{(c+d)k}{2}+(t-k+1)a-\frac{c+d}{2}t)/t = \lim_{t\to\infty}(\frac{2a-(c+d)}{2}t+\frac{(c+d)k}{2}+(1-k)a)/t = \frac{2a-(c+d)}{2} < 0$,[16] we know that the outcome by deviation is not preferred by player 2 and thus he has no incentive to infinitely deviate from his fairness strategy in this way. For the second deviating situation, since player 2 will return to play in his fairness strategy finally, the outcomes in the long run are the same with the non-deviating case. Since we are using the limit of means preference relation and only the outcomes in the long run are taken into consideration, we know that for player 2 the outcomes in this infinitely repeated game with deviation are indifferent to the outcomes without deviation and thus player 2 also has no incentive to deviate in this way. For the last deviating situation, suppose that player 2 always deviates for $m$ rounds and returns to adopt his fairness strategy for $n$ rounds, then player 2 can obtain the total payoff of $(k/2)*(c+d)+\frac{(t-k)}{m+n}(ma+(n/2)*(c+d))$ in $t$ rounds of the game. As $\lim_{t\to\infty}\frac{(k/2)*(c+d)+\frac{(t-k)}{m+n}(ma+(n/2)*(c+d))-\frac{c+d}{2}t}{t} = \frac{ma+nc/2+d}{m+n} - \frac{c+d}{2} < 0$,[17] the outcome by following this fairness strategy is more preferred by player 2 and thus he has no incentive to deviate in this way either.

If player 2 starts deviating from his fairness strategy when leaving state $(C, D)$, there are also three possible situations to consider. The first one is that player 2 deviates by choosing action $D$ and sticks to it forever; in this case the outcomes will always be $(D, D)$ thereafter, as indicated by the lowest self-loop on state $(D, D)$. The second situation is that player 2 deviates by choosing action $D$ first but takes action $C$ (transiting to state $(D, C)$) after staying in state $(D, D)$ for finite rounds only and finally resumes his fairness strategy forever (i.e., ending in the middle loop between state $(C, D)$ and $(D, C)$). The third situation is similar with the second one except that player 2 always alternates between deviating and resuming his fairness strategy endlessly in an alternate way, which can be represented by the lower half loop in the automaton. Based on similar analysis with previous case, the reader can verify that player 2 has no incentive to deviate from his fairness strategy in all these three situations.

Since the game is symmetric, we omit the proof when player 1 is assumed to unilaterally deviate from this fairness strategy. Therefore we get the conclusion that both players have no incentive to unilaterally deviate from their own fairness strategies and the pair of fairness strategies in Eqs. 3.12 and 3.13 is an FSE in the infinitely repeated PD game. □

[16] It is based on the known constraint of $2a < c + d$.

[17] The readers can verify it based on the known constraints of $c + d > 2a$ and $n \geq 2$. The reason of $n \geq 2$ is that it needs at least two steps to return to his fairness strategy for player 2 before deviating again (from state $(C, C)$ to state $(C, D)$ and then to state $(D, C)$).

#### 3.2.2.4 Discussion: Fairness Strategies Versus Trigger Strategies

In trigger strategies, players initially play a prior agreed action profile and punish the player whose behavior deviates from this agreement. The punishment degree of different trigger strategies can be quite different. However, in fairness strategy, there is no need for establishing a prior agreement between the players, and also the players do not need to observe their opponents' previous actions. The players' behaviors in fairness strategies are determined by their fairness attitudes, which are adaptively adjusted.

Trigger strategy can also be employed to achieve both highest and fair payoffs in both infinitely repeated AC and PD games with limit of means preference relation. For example, consider the infinitely repeated PD game in Fig. 3.16b. Players can make a prior agreement on playing $(D, C)$ and $(C, D)$ alternately and punish any player deviating from it forever by choosing $D$ (grim trigger). However, this is less flexible in handling the case when one player only deviates occasionally, since the deviating player will be punished forever and thus both players will receive the lowest payoff of 1 thereafter. In contrast, since fairness strategy does not explicitly require the players to detect and penalize the other player's deviation, it can tolerate this kind of unintentional mis-coordinated behavior. The player using fairness strategy can resume to cooperate with his opponent (by choosing $C$) as long as his payoff becomes higher than or equal with that of his opponent.

On the other hand, trigger strategy is also less flexible in that we need to design a specific trigger strategy for each game beforehand. For example, consider the PD game in Fig. 3.16b and the asymmetric game in Fig. 3.21 together. It is obvious that the previous trigger strategy for the PD game cannot be applied to this asymmetric game. In this game, it is required to achieve one round of $(D, C)$ and three rounds of $(C, D)$ alternately to resolve the conflict between achieving fairness and maximizing each player's payoff. In contrast, there exists a pair of fairness strategies in Eqs. 3.14 and 3.15 which is applicable to both games and also in fairness strategy equilibrium.

**Fig. 3.21** An example of asymmetric game

| Player 1's payoff, Player 2's payoff | | Player 2's action | |
|---|---|---|---|
| | | C | D |
| Player 1's action | C | 1.5,1.5 | 2,4 |
| | D | 6,0 | 1.5,1.5 |

It is easy to prove and we leave it to the readers. Overall we can see that fairness strategy provides more generality and flexibility compared with trigger strategy.

$$f_i^t = \begin{cases} \alpha_i^t = 0.5, \beta_i^t = 0 & \text{if } t = 1 \\ \alpha_i^t = \alpha_i^{t-1} - 2, \beta_i^t = \beta_i^{t-1} + 1.5 & \text{if } t > 1 \text{ and } cu_i^{t-1} > cu_j^{t-1} \text{ and } cu_i^{t-2} \leq cu_j^{t-2} \\ \alpha_i^t = \alpha_i^{t-1}, \beta_i^t = \beta_i^{t-1} & \text{if } t > 1 \text{ and } cu_i^{t-1} \leq cu_j^{t-1} \text{ and } cu_i^{t-2} \leq cu_j^{t-2} \\ \alpha_i^t = \alpha_i^{t-1}, \beta_i^t = \beta_i^{t-1} & \text{if } t > 1 \text{ and } cu_i^{t-1} > cu_j^{t-1} \text{ and } cu_i^{t-2} > cu_j^{t-2} \\ \alpha_i^t = \alpha_i^{t-1} + 2, \beta_i^t = \beta_i^{t-1} - 1.5 & \text{if } t > 1 \text{ and } cu_i^{t-1} \leq cu_j^{t-1} \text{ and } cu_i^{t-2} > cu_j^{t-2} \end{cases} \tag{3.14}$$

$$f_j^t = \begin{cases} \alpha_j^t = -0.5, \beta_j^t = 1.5 & \text{if } t = 1 \\ \alpha_j^t = \alpha_j^{t-1} - 1, \beta_i^t = \beta_j^{t-1} + 2 & \text{if } t > 1 \text{ and } cu_j^{t-1} \geq cu_i^{t-1} \text{ and } cu_j^{t-2} < cu_i^{t-2} \\ \alpha_j^t = \alpha_j^{t-1}, \beta_j^t = \beta_j^{t-1} & \text{if } t > 1 \text{ and } cu_j^{t-1} < cu_i^{t-1} \text{ and } cu_j^{t-2} < cu_i^{t-2} \\ \alpha_j^t = \alpha_j^{t-1}, \beta_j^t = \beta_j^{t-1} & \text{if } t > 1 \text{ and } cu_j^{t-1} \geq cu_i^{t-1} \text{ and } cu_j^{t-2} \geq cu_i^{t-2} \\ \alpha_j^t = \alpha_j^{t-1} + 1, \beta_j^t = \beta_j^{t-1} - 2 & \text{if } t > 1 \text{ and } cu_j^{t-1} < cu_i^{t-1} \text{ and } cu_j^{t-2} \geq cu_j^{t-2} \end{cases} \tag{3.15}$$

# References


1. Shoham Y, Powers R, Grenager T (2007) If multi-agent learning is the answer, what is the question? Artif Intell 171:365–377
2. Hao JY, Leung HF (2010) Strategy and fairness in repeated two-agent interaction. In: Proceedings of ICTAI'10, Arras, pp 3–6
3. Hao JY, Leung HF (2012) Incorporating fairness into infinitely repeated games with conflicting interests for conflicts elimination. In: Proceedings of ICTAI'12, Athens, pp 314–321
4. Hao JY, Leung HF (2012) Incorporating fairness into agent interactions modeled as two-player normal-form games. In: Proceedings of PRICAI'12, Kuching
5. Watkins CJCH, Dayan PD (1992) Q-learning. Mach Learn 3:279–292
6. Claus C, Boutilier C (1998) The dynamics of reinforcement learning in cooperative multiagent systems. In: Proceedings of AAAI'98, Madison, pp 746–752
7. Littman M (1994) Markov games as a framework for multi-agent reinforcement learning. In: Proceedings of ICML'94, New Brunswick, pp 322–328
8. Nowé A, Parent J, Verbeeck K (2001) Social agents playing a periodical policy. In: Proceedings of ECML'01, vol 2176, pp 382–393. Springer, Berlin/New York
9. Verbeeck K, Nowé A, Parent J, Tuyls K (2006) Exploring selfish reinforcement learning in repeated games with stochastic rewards. Auton Agents Multi-agent Syst 14:239–269
10. Fehr E, Schmidt KM (1999) A theory of fairness, competition and cooperation. Q J Econ 114:817–868
11. Kahneman D, Tversky A (1979) Prospect theory: an analysis of decison under risk. Econometrica 47(2):263–291
12. Gagne R (1985) The conditions of learning, 4th edn. Holt, Rinehart and Winston, New York
13. Chevaleyre Y, Dunne PE et al (2006) Issues in multiagent resource allocation. Informatica 30:3–31
14. Simon H (1972) Theories of bounded rationality. Decis Organ 1:161–176
15. Chevaleyre Y, Endriss U, Lang J, Maudet N (2007) A short introduction to computational social choice. SOFSEM 4362:51–69

16. Rabin M (1993) Incorporating fairness into game theory and economics. Am Econ Rev 83:1281–1302
17. Dawes RM, Thaleri RH (1988) Anomalies: cooperation. J Econ Perspect 2:187–198
18. Thaler RH (1985) Mental accounting and consumer choice. Mark Sci 4:199–214
19. Kahneman D, Knetsch JL, Thaler RH (1986) Fairness as a constraint on profit seeking: entitlements in the market. Am Econ Rev 76:728–741
20. Camerer C, Thaler RH (1995) Ultimatums, dictators, and manners. J Econ Perspect 9:209–219
21. Agell J, Lundberg P (1995) Theories of pay and unemployment: survery evidence from swedish manufacturing firms. Scand J Econ XCVII:295–308
22. Bewley T (1995) A depressed labor market as explained by participants. Am Econ Rev Pap Proc LXXXV:250–254
23. Leventhal G, Anderson D (1970) Self-interest and the maintenance of equity. J Personal Soc Psychol 15:57–62
24. Geanakoplos J, Pearce D, Stacchetti E (1989) Psychological games and sequential rationality. Games Econ Behav 1:60–79
25. Bolton GE, Ockenfels A (2000) Erc-a theory of equity, reciprocity and competition. Am Econ Rev 90:166–193
26. Falka A, Fischbache U (2006) A theory of reciprocity. Games Econ Behav 54:293–315
27. Shoham Y, Power WR, Grenager T (2007) If multi-agent learning is the answer, what is the question? Artif Intell 171(7):365–377
28. Blount S (1995) When social outcomes aren't fair: the effects of causual attributions on preferences. Organ Behav Hum Decis Process LXIII:131–144

# Chapter 4
# Social Optimality in Cooperative Multiagent Systems

In this chapter, we turn to look at another important solution concept called *social optimality* which targets at maximizing the sum of all agents' payoffs involved. A socially optimal outcome is desirable in that it is not only optimal from the system-level's perspective but also Pareto optimal. To achieve socially optimal outcomes in cooperative environments, the major challenge is how each agent can coordinate effectively with others given limited information, since the behaviors of other agents coexisting in the system may significantly impede the coordination process among them. The coordination problem becomes more difficult when the environment is uncertain (or stochastic) and each agent can only interact with its local partners if we consider a topology-based interaction environment [1, 6].

We start with the multiagent coordination problems within a population of agents where the agents share common interests and the same reward function: the increase in individual satisfaction also results in the increase in the satisfaction of the group. We propose two different types of learners (individual action learner and joint action learner) within the social learning framework, which can be considered as the parallel extension of the two types of learners (independent learner and joint action learner) from repeated interaction framework to the social learning framework [2, 3] (Sect. 4.1). Following that, we extend our learning framework to the context of general-sum games where agents' payoff is not necessarily always the same. The effectiveness of this framework can be verified from extensively experimental simulations comparing with previous work [4, 5] (Sect. 4.2). Lastly, we turn to look at a particular type of multiagent interaction environment—multiagent negotiation problem, where multiple agents decide the allocation of a bundle of resources through negotiation (Sect. 4.3). We introduce a state-of-the-art distributed negotiation protocol through which the agents are able to achieve socially optimal negotiation outcome eventually.



J. Hao, H.-f. Leung, *Interactions in Multiagent Systems: Fairness, Social Optimality and Individual Rationality*, DOI 10.1007/978-3-662-49470-7_4

## 4.1 Reinforcement Social Learning of Coordination in Cooperative Games

In cooperative games, the agents share common interests and the same reward function; the increase in individual satisfaction also results in the increase in the satisfaction of the group. A commonly adopted learning framework for studying the coordination problem within cooperative games is to consider two (or more) players playing a repeated (stochastic) game, in which the agents learn their optimal policies toward coordination through repeated interactions [6]. There are a number of challenges the agents have to face when learning in cooperative games. One major difficulty is the *equilibrium selection problem* [7], i.e., multiple optimal joint actions exist under which coordination between the agents is required in selecting among multiple optimal joint actions. Another issue is the *stochasticity problem* [8], i.e., the game itself can be nondeterministic. In this case, the challenge is that the agents need to distinguish whether the different payoffs received by performing an action are caused by the explorations of the other agent or the stochasticity of the game itself.

Until now, various multiagent reinforcement learning algorithms [6, 9–15] have been proposed in the literature to solve the coordination problem in cooperative games. Most of previous works heavily rely on the $Q$-learning algorithm [16] as the basis and can be considered as various modifications of $Q$-learning algorithms to achieve better coordination on optimal joint action(s) in cooperative games. Much of the previous work adopts repeated games and stochastic games to model the interactions among the learning agents, and the typical goal is to learn the policy of reaching optimal joint action(s) finally. However, in practical complex systems, the interactions between agents can be sparse [17], i.e., it is highly likely that each agent may not have the opportunity to always interact with the same partner, and its interaction partners may change frequently and randomly. In this situation, an agent's policy that achieved coordination on an optimal joint action with one partner may fail when it comes to a different partner next time. Each agent learns its policy through repeated interactions with different opponents. This kind of learning is termed as *social learning* [18] to distinguish from the case of learning from repeated interactions with the same partner(s) [19]. Previous work [18] has investigated the emergence of consistent coordination policies (norms) in conflicting-interest games under this social learning framework; however, little work has been done in studying the coordination problem in cooperative environments under such a learning framework. Apart from the difficulties previously mentioned in the cooperative environments where the interaction partners are fixed, achieving coordination on optimal joint actions under the social learning framework can be more challenging due to the additional stochasticity caused by non-fixed interaction partners. It is not clear a priori if all the agents can still learn to converge to a consistent optimal coordination policy in such a situation. To this end, we study the multiagent coordination problem in cooperative multiagent environments under the social learning framework, which will be introduced in Sect. 4.1.1.

**Algorithm 2** Overall interaction protocol of the social learning framework

```
1: for a number of rounds do
2:     repeat
3:         two agents are randomly chosen from the population, and one of them is assigned as the
           row player and the other one as the column player.
4:         both agents play a two-player cooperative game by choosing their actions independently
           and simultaneously.
5:     until all agents in the population have been selected
6:     for each agent in the population do
7:         update its policy based on its experience in the current round
8:     end for
9: end for
```

### *4.1.1 Social Learning Framework*

Under the social learning framework, there are a population of $N$ agents, and each agent learns its policy through repeated pairwise interactions with the rest of agents in the population. The interaction between each pair of agents is modeled as a two-player cooperative game. During each round, each agent interacts with a randomly chosen agent from the population, and one agent is randomly assigned as the row player and the other agent as the column player. The agents are assumed to know their roles, i.e., either as row player or column player, during each interaction. After each interaction, each agent updates its policy based on the learning experience it receives from the interaction. The overall interaction protocol under the social learning framework is presented in Algorithm 2.

In the two-player repeated interaction learning framework [9], two classes of learners (independent learner and joint action learner) are distinguished in terms of the amount of information that an agent can perceive from each interaction: (1) its own payoff and action alone and (2) its own payoff, action, and its interaction partner's action choice. Under the social learning framework, two different types of settings can also be distinguished depending on the amount of information each agent can have access to, which will be introduced in detail in Sect. 4.1.1.1. Besides, each agent needs to employ a learning strategy to make its decisions under each setting, which will be described in Sect. 4.1.1.2.

#### 4.1.1.1 Observation

In each round, each agent is only allowed to interact with one other agent randomly chosen within the population. We define each pair of interacting agents as being in the same group. For the traditional setting involving only two (or more) fixed agents repeatedly playing a cooperative game (within the same group), the learning experience of each agent comes from its own group only. However, under the social learning framework, since all agents interact with their own interaction partners concurrently, different agents may be exposed to experience of those agents from other groups (their actions and payoffs). One extreme case would be that each agent can have access to the experience of all the other agents in the population (global observation), which is the assumption commonly adopted under the evolutionary game-theoretic framework [4, 20]. However, this could be unrealistic in practical interaction scenarios due to communication limitation. Allowing agents to observe the information of other agents outside their direct interactions may generally result in a faster learning rate and facilitate coordinating on optimal solutions. To balance the trade-off between global observations and local interactions, here we allow each agent to have access to the information of other $M$ groups randomly chosen in the system at the end of each round. If the value of $M$ is equal to the total number of groups existing in the population, it becomes equivalent with the case of global observation; if the value of $M$ is zero, it reduces to the case of local interactions. The value of $M$ represents the connectivity degree in terms of information sharing among different groups. By spreading this information into the population with $M \ll N$, it serves as an alternative biased exploration mechanism to accelerate the agents' convergence to optimal joint actions while keeping communication overhead at a low level at the same time. Besides, it has been found that reducing the amount of information available to rational decision-makers ($M \ll N$) can be an effective mechanism for achieving system stability and improving system-level performance [21].

Similar to the two settings adopted in the case of two-agent repeated interactions [9], here we also identify two different settings depending on the amount of information that each agent has access to during each interaction. In the first setting, apart from its own action and payoff, each agent can also observe the actions and payoffs of all agents with the same role (row player or column player) as itself from other $M$ groups. Formally, the information that each agent $i$ can perceive at the end of each round $t$ can be represented as the set $S_i^t = \{\langle a_i^t, r^t\rangle, \langle b_1^t, r_1^t\rangle, \langle b_2^t, r_2^t\rangle, \ldots, \langle b_M^t, r_M^t\rangle\}$. Here $\langle a_i^t, r^t\rangle$ is the action and payoff of agent $i$ itself, and the rest are the actions and payoffs of those agents with the same role from other $M$ groups in the system.

For the second setting, apart from the same information available in the first setting, each agent is also able to perceive the action choices of its interaction partner and also those agents with opposite role from other $M$ groups. Formally, the experience for each agent $i$ at the end of each round $t$ can be denoted as the set $P_i^t = \{\langle (a_i^t, a_j^t), r^t\rangle, \langle (b_1^t, c_1^t), r_1^t\rangle, \langle (b_2^t, c_2^t), r_2^t\rangle, \ldots, \langle (b_M^t, c_M^t), r_M^t\rangle\}$. Here $\langle (a_i^t, a_j^t), r^t\rangle$ is the joint action and payoff of agent $i$ and its interaction partner, and the rest are the joint actions and payoffs information of other $M$ groups.

The learning strategy of each agent can be quite different depending on the settings that the agents are situated in, and we distinguish two types of learners, *individual action learner* and *joint action learner*, which correspond to the above two settings respectively and will be described in next section.

#### 4.1.1.2 Learning Strategy

In general, to achieve coordination on optimal joint actions, an agent's policies as the row or column player can be the same or different depending on the characteristics of the games. Accordingly, we propose that each agent has a pair of strategies, one used as the row player and the other used as the column player, to play with any other agent from the population. The strategy we develop here is based on $Q$-learning technique [16] similar to previous work [9, 11–13]. Depending on the amount of information that the agents are assumed to have access to, there are two distinct ways of applying $Q$-learning to the social learning framework.

Individual Action Learner

In the first setting, each agent can be considered as being unaware of the existence of its interacting partner and the interacting partners of those agents with the same role from other $M$ groups. Naturally each agent holds a $Q$-value $Q(s, a)$ for each action $a$ under each state $s \in \{Row, Column\}$, which keeps record of action $a$'s past performance and serves as the basis for making decisions. Each agent has two possible states corresponding to its roles as the row player or column player. We assume that each agent knows its current role (row player or column player) during each interaction. At the end of each round $t$, each agent $i$ picks an action (randomly choosing in case of a tie) with the highest payoff from the set $S_i^t$ and updates its $Q$-values of these actions following Eq. 4.1,

$$Q_i^{t+1}(s,a) = Q_i^t(s,a) + \alpha^t(s)[r(a) * \text{freq}(a) - Q_i^t(s,a)] \tag{4.1}$$

where $r(a)$ is the highest payoff of action $a$ among all elements in set $S_i^t$, freq$(a)$ is the frequency that action $a$ occurs with the highest reward $r(a)$ among the set $S_i^t$, and $\alpha^t(s)$ is the current learning rate in state $s$.

Intuitively, the above update rule incorporates both the optimistic assumption and the FMQ heuristic. On one hand, this update rule is optimistic since we only update the $Q$-values of those actions that receive the highest payoff based on the current round's experience. On the other hand, similar to the FMQ heuristic, the update rule also takes into consideration the information of how frequent each action $a$ can receive the highest payoff based on the current round's experience.

Each agent chooses its action based on the corresponding set of $Q$-values during each interaction according to the $\epsilon$-greedy mechanism, which is commonly adopted for performing biased action selection in reinforcement learning literature. Under this mechanism each agent chooses its action with the highest $Q$-value with probability $1 - \epsilon$ to exploit the action with best performance currently (random selection in case of a tie) and makes random choices with probability $\epsilon$ for the purpose of exploring new actions with potentially better performance.

Joint Action Learner

In the second setting, each agent can have access to the joint actions of its own group and other $M$ groups as well. Accordingly, each agent can learn the $Q$-values for each joint action in contrast to learning $Q$-values for individual actions only in the first setting. Specifically, at the end of each round $t$, each agent $i$ updates its $Q$-values for each joint action $\overrightarrow{a}$ belonging to the set $S^t$ as follows:

$$Q_i^{t+1}(s, \overrightarrow{a}) = Q_i^t(s, \overrightarrow{a}) + \alpha^t(s)\left[r(\overrightarrow{a}) - Q_i^t(s, \overrightarrow{a})\right] \tag{4.2}$$

where $r(\overrightarrow{a})$ is agent $i$'s payoff under the joint action $\overrightarrow{a}$ and $\alpha^t(s)$ is its current learning rate under state $s$.

After enough explorations, we can see that the above $Q$-values can reflect the expected performance of each joint action in the system, but each agent still needs to determine the relative performance of its individual actions to make decisions. For each agent, the performance of its individual actions depends crucially on the action choices of its interaction partners, which can be estimated based on the past experience $P_i^t$.

At the end of each round $t$, for each action $a$, let us first define $P_a(s) = \{r \mid \langle(a, b), r\rangle \in P_i^t\}$, $s \in \{Row, Column\}$ and $r_a^{\max} = \max\{P_a(s)\}$. The value of $r_a^{\max}$ reflects the maximum possible payoff that agent $i$ can obtain from performing action $a$ under state $s$ (i.e., row player or column player). However, agent $i$'s actual expected payoff of performing action $a$ also largely depends on the action choices of its interaction partners. To take this factor into consideration, each agent $i$ assesses the relative performance $EV(s, a)$ of each individual action $a$ as follows:

$$EV(s, a) = r_a^{\max} \times freq(a, b) \tag{4.3}$$

where freq$(a, b)$ represents agent $i$'s belief of the frequency of its interaction partners performing action $b$ based on current round experience $P_i^t$, where the joint action pair $(a, b)$ corresponds to the maximum payoff $r_a^{\max}$ in $P_i^t$.

Similar to the case of IALs, the way of evaluating the relative performance of each individual action $a$ incorporates not only the optimistic assumption but also the information of the frequency that the maximum payoff can be received by performing action $a$. However, for the JALs, for each action $a$ they are able to identify the matching action of their interaction partners corresponding to the maximum payoff, and thus it is expected that they can get more accurate estimations of the relative performance of each individual action.

Each agent chooses its action based on the EV-values $EV(s, .)$ of its individual actions in the same way as it would use $Q$-values for IALs following $\epsilon$-greedy mechanism. Specifically, each agent chooses an action $a$ with the highest EV-value $EV(s, a)$ with probability $1 - \epsilon$ to exploit the action with best performance currently (random selection in case of a tie) and makes random choices with probability $\epsilon$ for the purpose of exploring new actions with potentially better performance.

### 4.1.2 Experimental Evaluations

In this section, we first evaluate the learning performance of both IALs and JALs under the social learning framework. All the experiments are conducted in a population of 100 agents, and the value of $M$ is set to 5. Two representative deterministic cooperative games (the climbing game and the penalty game) are considered, which were first proposed in [9] and have been commonly adopted as the test beds in previous work [9, 11–13]. Two variants of the climbing games discussed in [12] are also considered here, partially stochastic climbing game and fully stochastic climbing game, to evaluate the learning performance in stochastic environments. To better understand the benefits of learning from the experience of other groups, we also further investigate the influences of the value of $M$ on the coordination performance of the agents.

#### 4.1.2.1 Performance Evaluation

Deterministic Games

We consider two particularly difficult coordination problems: the climbing game (Fig. 4.1a) and the penalty game (Fig. 4.1b). The climbing game has one optimal joint action $(a, a)$ and two joint actions $(a, b)$ and $(b, a)$ with high penalties. Figure 4.2 shows the percentage of agents reaching the optimal joint action $(a, a)$ of the climbing game as a function of the number of rounds for both IALs and JALs under the social learning framework. We can see that both IALs and JALs can successfully learn to coordinate on the optimal joint action $(a, a)$ without significant difference. Since there is no stochasticity in the climbing game, there is no significant advantage for the JALs by having the additional joint action information at their disposal. Similar results that there is no significant performance difference between IALs and JALs in the setting of two-player repeated cooperative games are also found in previous work [9].

(a)

| 1's payoff 2's payoff | | Agent 2 | | |
|---|---|---|---|---|
| | | a | b | c |
| Agent 1 | a | 11 | -30 | 0 |
| | b | -30 | 7 | 6 |
| | c | 0 | 0 | 5 |

(b)

| 1's payoff 2's payoff | | Agent 2 | | |
|---|---|---|---|---|
| | | a | b | c |
| Agent 1 | a | 10 | 0 | k |
| | b | 0 | 2 | 0 |
| | c | k | 0 | 10 |

**Fig. 4.1** Payoff matrices for the climbing game and the penalty game. (**a**) The climbing game. (**b**) The penalty game

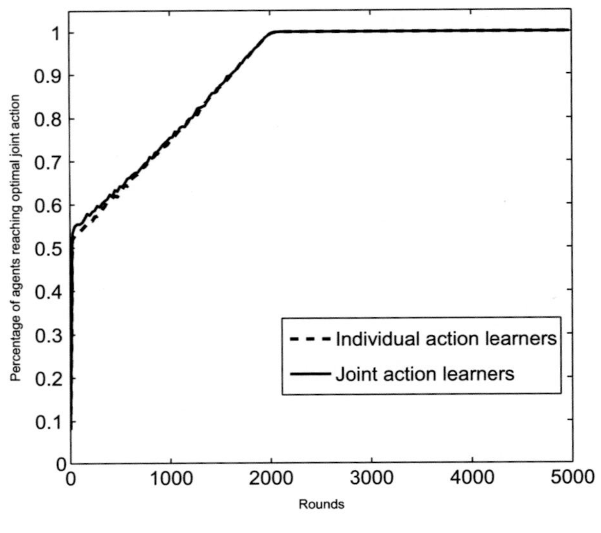


**Fig. 4.2** Percentage of agents coordinating on optimal joint action $(a, a)$ in each round (averaged over 500 times)

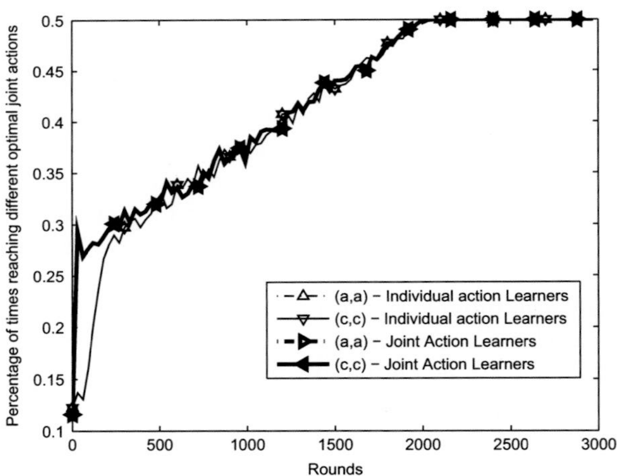


**Fig. 4.3** Percentage of agents coordinating on optimal joint actions $(a, a)$ or $(c, c)$ in each round (averaged over 500 times)

In the penalty game, apart from the existence of two joint actions with heavy penalties $(a, c)$ and $(c, a)$ $(k = -50)$, there also exist two optimal joint actions $(a, a)$ and $(c, c)$ in contrast to only one optimal joint action in the climbing game. Thus it is more difficult for the agents to reach an optimal joint action in the penalty game since the agents also need to agree on which optimal joint action to choose. Figure 4.3 shows the percentage of agents reaching the optimal joint actions $(a, a)$ or $(c, c)$ of the penalty game as a function of the number of rounds for both IALs and JALs under the social learning framework. We can observe that all agents in the system can successfully learn to converge on optimal joint actions after around 2000 rounds for both IALs and JALs. For both IALs and JALs, on average half of times, all agents learn to adopt the policy of coordinating on the optimal joint

(a)

| 1's payoff 2's payoff | | Agent 2 | | |
|---|---|---|---|---|
| | | a | b | c |
| Agent 1 | a | 11 | -30 | 0 |
| | b | -30 | 14/0 | 6 |
| | c | 0 | 0 | 5 |

(b)

| 1's payoff 2's payoff | | Agent 2 | | |
|---|---|---|---|---|
| | | a | b | c |
| Agent 1 | a | 10/12 | 5/-65 | 8/-8 |
| | b | 5/-65 | 14/0 | 12/0 |
| | c | 5/-5 | 5/-5 | 10/0 |

**Fig. 4.4** Payoff matrices for the partially stochastic climbing game and the fully stochastic climbing game. (**a**) The partially stochastic climbing game (the joint action $(b, b)$ yields the payoff of 14 or 0 with equal probability). (**b**) The fully stochastic climbing game (each joint action yields two different payoffs with equal probability)

action $(a, a)$ and learn to converge to another optimal joint action $(c, c)$ the other half of times. Besides, similar to the climbing game, since the penalty game is also deterministic, there is no significant difference in the learning performance between IALs and JALs even though the JALs have more information at their disposal.

Partially Stochastic Climbing Game

In this section, we consider the first variant of the climbing game—the partially stochastic climbing game (shown in Fig. 4.4a) [11]. This version of the climbing game is different from the original one in that a stochastic payoff is introduced for the joint action $(b, b)$. The joint action $(b, b)$ yields the payoff of 14 and 0 with probability 0.5. This partially stochastic climbing game is in essence equivalent with the original climbing game, since the expected payoff that each agent receives by achieving the joint action $(b, b)$ remains the same with the payoff in the original deterministic one.

Figure 4.5a illustrates the dynamics of proportion of agents coordinating on the optimal joint action $(a, a)$ in each round for both the cases of IALs and JALs. First, we can see that both IALs and JALs can reach full coordination on the optimal joint action $(a, a)$ after approximately 2200 rounds. Another observation is that the JALs do perform significantly better than that of IALs in terms of the percentage of agents coordinating on $(a, a)$ after a short number of rounds. This is expected since the JALs can distinguish the $Q$-values of different joint actions and have the ability of quickly identifying which action pair is optimal even though the game is partially stochastic. In contrast, for the IALs, since they cannot perceive the actions of their interaction partners, they cannot distinguish between the noise from the stochasticity of the game and the explorations of their interaction partners. Thus it is more difficult for the IALs to learn the actual $Q$-values of their individual actions.

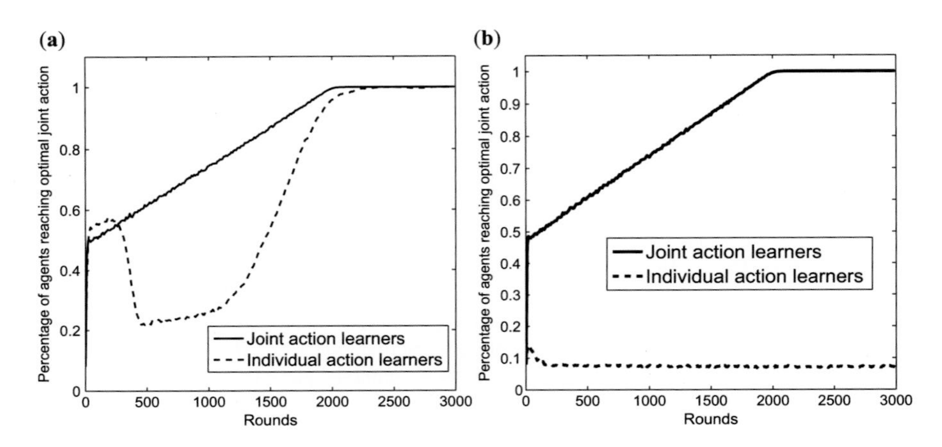


**Fig. 4.5** Percentage of agents coordinating on optimal joint actions $(a, a)$ in each round (averaged over 500 times) in partially stochastic climbing game and fully stochastic climbing game. (**a**) Partially stochastic climbing game. (**b**) Fully stochastic climbing game

### Fully Stochastic Climbing Game

Next let us further consider another variant of the climbing game—the fully stochastic climbing game (shown in Fig. 4.4b) [11]. The characteristic of this game is that all joint actions yield two different payoffs with equal probability. This version of the climbing game is functionally equivalent with the originally deterministic one, since the average payoff of each joint action remains the same with the payoff in the originally deterministic version. It is worth mentioning that, under this fully stochastic climbing game, previous work [8] shows that both (recursive) FMQ heuristic [11, 12] and optimistic assumption [11] fail to coordinate on the optimal joint action $(a, a)$, and the lenient learners [13] only can coordinate on the optimal joint action $(a, a)$ in approximately 90 % of the runs on average.

Figure 4.5b illustrates the percentage of agents coordinating on the optimal joint action $(a, a)$ of the fully stochastic climbing game as a function of the number of rounds for both IALs and JALs under the social learning framework. From Fig. 4.5b, we can see that JALs can always successfully learn to coordinate on optimal joint action $(a, a)$, while the IALs fail to do that. Even though the IALs take into consideration the information of the occurrence frequency of the action corresponding to the highest payoff, they cannot accurately learn the $Q$-values of each action due to too much noise introduced by the full stochasticity of the game. However, for the JALs, since they have more information at their disposal in terms of joint $Q$-values and the occurrence frequency of each action of their interaction partners, they are able to get a better estimation of the relative performance of each individual action (EV-values). Indeed we can see that full coordination on optimal joint action among JALs can be achieved.

#### 4.1.2.2 Effects of Degree of Information Sharing Degree Among Agents

One major difference between the social learning framework and the commonly adopted setting of repeated games is that the agents may have more information (other pairs of interaction agents in the population) at their disposal when they learn their coordination policies. Therefore, in this section, we evaluate the influences of the amount of information shared among agents (the value of $M$) in the social learning framework on the learning performance of the agents.

Figure 4.6 shows the learning dynamics of the percentages of IALs coordinating on optimal joint action with different values of $M$ in the partially stochastic climbing game (see Fig. 4.4a). We can see that when the value of $M$ is too small, the agents fail to learn a consistent policy of coordinating on the optimal joint action. As the value of $M$ increases ($2 \rightarrow 4 \rightarrow 6$), better performance can be achieved in terms of the percentages of agents coordinating on the optimal joint action. However, it is interesting to notice that the coordination performance gradually decreased when the value of $M$ is further increased ($6 \rightarrow 8 \rightarrow 10$). This implies that either giving too little or too much information available to each agent can impede efficient coordinations among agents when the agents learn based on local information only. In the partially stochastic climbing game, the IALs need to learn to distinguish between the noise from the stochasticity of the game and the explorations of their interaction partners. If the value of $M$ is too small, the agents cannot get an accurate estimation of the performance of each individual action since the sampling size is too small. However, if the value of $M$ becomes too large, the noise introduced by the explorations of the interaction partners will dominate, and the stochasticity of the game will be neglected. Accordingly, the IALs will also get a biased estimation of the performance of each action when $M$ is too large. On the other hand, when the climbing game is deterministic, the above problem does not exist; thus, the learning

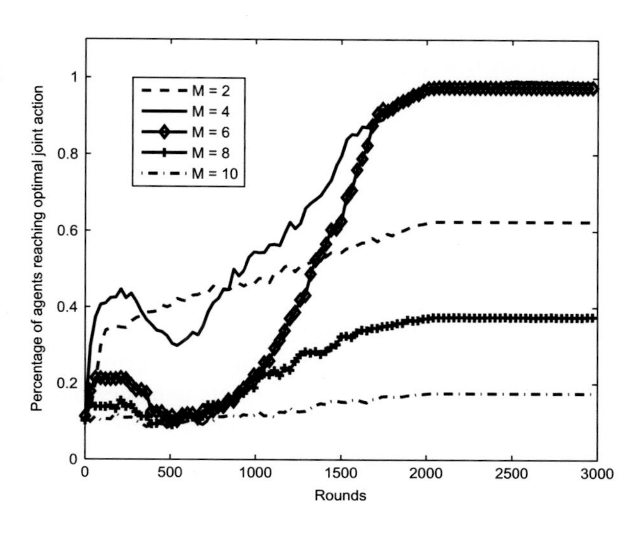


**Fig. 4.6** Percentage of IALs coordinating on optimal joint actions $(a, a)$ in each round (averaged over 500 times)

performance of IALs is always improved as the value of $M$ is increased when the value $M$ is small, and also the overall performance will not be decreased with the value of $M$ further increased. The results are omitted due to space constraints.

For JALs, since they can perceive the joint actions of all the $M$ pairs of interaction agents, the larger $M$ is, the more accurate estimations of the expected payoff of each joint action ($Q(a, b)$) they can get. Given a stochastic game, the JALs actually learn over its corresponding deterministic version in which the payoff of each joint action is replaced by its expected value. Therefore, similar to the analysis of IALs in deterministic games, there are no detrimental effects on the learning performance of JALs when the value of $M$ is increased. Due to space constraints, the results are not shown here.

## 4.2 Reinforcement Social Learning of Coordination in General-Sum Games

Next we extend the social learning framework for cooperative games in previous section to the context of general-sum games. We consider the problem of how a population of agents is able to effectively learn to coordinate on socially optimal solutions in general-sum games within social learning framework. It is worth noting that the desirable coordination outcomes in cooperative games in Sect. 4.1 are always socially optimal, which can be considered as a special case of coordination in general-sum games. Compared with previous work [4, 20, 22], two major novel contributions of this framework are as follows. First, the *observation* mechanism is introduced to reduce the amount of communication required among agents. Second, the agents' learning strategies are based on reinforcement learning technique instead of evolutionary learning. Each agent explicitly keeps the record of its current state in its learning strategy and learns its optimal policy for each state independently. In this way, the learning performance is much more stable, and also it is suitable for both symmetric and asymmetric games.

### *4.2.1 Social Learning Framework*

#### 4.2.1.1 Interaction Protocol

We consider a setting involving a population of agents in which each agent interacts with another agent randomly chosen from the population each round. The interaction between each pair of agents is modeled as a single-stage normal-form game, and the payoff each agent receives in each round only depends on the joint action of itself and its interaction partner. Under this learning framework, each agent learns its policy through repeated interactions with multiple agents, which is termed as *social learning* [18], in contrast to the case of learning from repeated interactions

| 1's payoff, 2's payoff | | Player 2's action | |
|---|---|---|---|
| | | C | D |
| Player 1's action | C | 0, 0 | 1, 2 |
| | D | 2, 1 | 0, 0 |

**Fig. 4.7** An instance of the anti-coordination game

1: **for** a number of rounds **do**
2: **repeat**
3: two agents are randomly chosen from the population, and one of them is assigned as the row player and the other one as the column player.
4: both agents play a two-player normal-form game and receive their corresponding payoffs
5: **until** all agents in the population have been selected
6: **for** each agent in the population **do**
7: update its policy based on its past experience
8: **end for**
9: **end for**

**Fig. 4.8** Overall interaction protocol of the social learning framework

against the same opponent in the two-agent case. We aim at solving the problem of how the agents are able to learn to coordinate on socially optimal outcomes in general-sum games through *social learning*.

The overall interaction protocol under the social learning framework is presented in Fig. 4.8. During each round, each agent interacts with another agent randomly chosen from the population. Since we consider in the context of general-sum games, during each interaction one agent is randomly assigned as the row player and the other agent as the column player. At the end of each round, each agent updates its policy based on the learning experience it receives so far.

#### 4.2.1.2 Observation

In each round, each agent is only allowed to interact with one other agent randomly chosen from the population, and each pair of interacting agents is regarded as being in the same group. On one hand, during each (local) interaction, each agent typically can perceive its own group's information (i.e., the action choices and payoffs of itself and its interaction partner), which is also known as the perfect monitoring assumption [15]. This assumption is commonly adopted in previous work in two-player repeated interaction settings. On the other hand, in the social learning framework, different from the two-player interaction scenario, different agents may be exposed to information of agents from other groups. Allowing agents to observe the information of other agents outside their direct interactions may result in faster learning rate and facilitate coordinating on optimal solutions. In previous work [4, 20], it is assumed that each agent can have access to all the other agents' information (their action choices and payoffs) in the system (global observation), which may be unpractical in realistic environments due to the high communication cost involved. To balance the trade-off between global observations and local interactions, here we only allow each agent to have access to the information of other $M$ groups randomly chosen in the system at the end of each round. If the value of $M$ is equal to the number of agents, it becomes equivalent with the case of allowing global observation; if the value of $M$ is zero, it reduces to the case of local interactions. The value of $M$ represents the connectivity degree in terms of information sharing among different groups. By spreading this information into the population, it serves as an alternative biased exploration mechanism to accelerate the agents' convergence to the desired outcomes while keeping communication overhead at a low level. Besides, it has been found that reducing the amount of information available to rational decision-makers (letting $M \ll N$) can be an effective mechanism for achieving system stability and improving system-level performance [21].

#### 4.2.1.3 Learning Strategy

When an agent interacts with its interacting partner in each round, it needs to employ a learning strategy to make its decisions. Since in general-sum games the payoff matrices of the row and column players are usually different, each agent may need to behave differently according to its current role (either as the row player or column player). To this end, we assume that each agent has a pair of strategies, one used as the row player and the other used as the column player, to play with any other agent from the population. The strategy we develop here is based on $Q$-learning technique [16], in which each agent holds a $Q$-value $Q(s, a)$ for each action $a$ in each state $s \in \{Row, Column\}$, which keeps record of the action's past performance and serves as the basis for making decisions. We assume that each agent knows its current role during each interaction. Accordingly, each agent chooses its action based on the corresponding set of $Q$-values of its current role during each interaction.

##### How to Select the Action

Each agent chooses its action based on $\epsilon$-greedy mechanism, which is commonly adopted for performing biased action selection in reinforcement learning literature. Under this mechanism each agent chooses its action with the highest $Q$-value with probability $1 - \epsilon$ for exploiting the action with best performance currently (random selection in case of a tie) and makes random choices with probability $\epsilon$ for the purpose of exploring new actions with potentially better performance.

### How to Update the $Q$-Values

According to Sect. 4.2.1.2, we know that in each round the information that each agent has access to consists of its own group's information and the information of other $M$ groups randomly chosen from the population. The traditionally and commonly adopted approach for a selfish agent is to use its individual payoff received from performing an action as the reinforcement signal to update its strategy ($Q$-values). However, since an agent is situated in a social context, its selfish behaviors can have implicit and detrimental effects on the behaviors of other agents coexisting in the environment, which may in return lead to undesirable results for both individual agents and the overall system. Accordingly, an alternative notion proposed for solving the above problem is *social rationality* [23, 24], under which each agent evaluates the performance of each action in terms of its worth to the overall group instead of itself. We consider an agent as an individually rational agent if its strategy is updated based on its own payoff only, and this kind of update scheme is called *individually rational update*, while we consider an agent as a socially rational entity if its strategy is updated based on the sum of all the agents' payoffs in the same group, and this kind of update scheme is called *socially rational update*.

Adopting which way to update the agents' strategies is better depends on the type of games being played. For cooperative games, it is indifferent between *socially rational update* and *individually rational update*, since the agents share the common interests; for competitive (zero-sum) games, it is meaningless to perform *socially rational update* since the sum of both players' payoffs under every outcome is always equal to zero and it is more reasonable for the agents to update their strategies based on their individual payoffs only; for mixed games, it can be beneficial for the agents to update their strategies in the socially rational manner, which can facilitate the agents to coordinate on win-win outcomes rather than loss-loss outcomes (e.g., the prisoner's dilemma game). One the other hand, it may be better to adopt individually rational update to prevent malicious exploitations when interacting against individually rational agents.

Since we assume that the agents have no prior information of the game they are playing, each agent should learn to determine the appropriate way of updating its strategy adaptively, i.e., updating its strategy based on either its individual payoff or the joint payoff(s) of the group(s) or the combination of both. To handle the individually/socially rational update selection problem, we propose that each agent $i$ is equipped with two weighting factors $w_{i,il}$ and $w_{i,sl}$ representing its individual rational and socially rational degrees satisfying the constraint $w_{i,il} + w_{i,sl} = 1$. In general each agent's $Q$-values are updated in the individually rational and socially rational manner as follows:

$$Q_i^{t+1}(s,a) = Q_i^t(s,a) + \alpha^t(s)w_{i,il}^t(r_{i,il}(s,a) - Q_i^t(s,a)),\ s \in \{Row, Column\} \tag{4.4}$$

$$Q_i^{t+1}(s,a) = Q_i^t(s,a) + \alpha^t(s)w_{i,sl}^t(r_{i,sl}(s,a) - Q_i^t(s,a)),\ s \in \{Row, Column\} \tag{4.5}$$

**Algorithm 3** Adaptive weighting factors updating strategy for agent $i$

1: initialize the value of $w_{i,il}$ and $w_{i,sl}$.
2: initialize the updating direction $UD^0$, i.e., towards increasing the weight on individual learning or social learning.
3: **for** at the end of each period T **do**
4: keep record of its own performance (e.g.,the average payoff $avg^T$) within this period.[1]
5: **if** it is not the first adjustment period ($T > 1$) **then**
6: **if** $avg^T > avg^{T-1}$ or $\frac{|avg^T - avg^{T-1}|}{\max(avg^T, avg^{T-1})} \leq$ Threshold **then**
7: keep the updating direction $UD^T$ unchanged
8: **else**
9: reverse the updating direction $UD^T$
10: **end if**
11: **if** $UD^T$ equals towards social learning **then**
12: $w_{i,sl} = w_{i,sl} + \delta$
13: $w_{i,il} = w_{i,il} - \delta$
14: **else**
15: $w_{i,il} = w_{i,il} + \delta$
16: $w_{i,sl} = w_{i,sl} - \delta$
17: **end if**
18: **else**
19: **if** $UD^T$ equals towards social learning **then**
20: $w_{i,sl} = w_{i,sl} + \delta$
21: $w_{i,il} = w_{i,il} - \delta$
22: **else**
23: $w_{i,il} = w_{i,il} + \delta$
24: $w_{i,sl} = w_{i,sl} - \delta$
25: **end if**
26: **end if**
27: **end for**

Here $Q_i^t(s, a)$ is agent $i$'s $Q$-function on action $a$ at time step $t$ in state $s$, $\alpha^t(s)$ is the learning rate in state $s$ at time step $t$, $r_{i,il}(s, a)$ is the payoff agent $i$ uses to update its $Q$-value of action $a$ in the individually rational manner, and $r_{i,sl}(s, a)$ is the payoff agent $i$ uses to update its $Q$-value in the socially rational manner. We will discuss and introduce how to determine the values of $r_{i,il}(s, a)$ and $r_{i,sl}(s, a)$ later. Notice that each agent shares the same set of weighting factors $w_{i,il}$ and $w_{i,sl}$ no matter whether it is the row player or the column player. The value of $\alpha^t(s)$ is decreased gradually: $\alpha^t(s) = \frac{1}{No.(s)}$, where $No.(s)$ is the number of times that state $s$ has been visited until time $t$.

In order to optimize its own performance in different environments, the values of each agent's weighting factors should be able to be adaptively updated based on its own past performance. Therefore, the natural question is how to update the values of $w_{i,il}$ and $w_{i,sl}$ for each agent. Here we propose an adaptive strategy for agents to make their decisions shown in Algorithm 3. Each agent adaptively updates its two

[1]In the current implementation, we only take each agent's payoff during the second-half period into consideration in order to get a more accurate evaluation of the actual performance of the period.

weighting factors in a greedy way based on its own past performance periodically. To update its weighting factors, each agent maintains an updating direction, i.e., toward either increasing the weight on individual rational update or the weight on socially rational update. We consider a fixed number $m$ of steps as a period. At the end of each period, each agent $i$ will adaptively adjust these two weighting factors by $\delta$ following the current updating direction, where $\delta$ is the adjustment degree. For example, if the current updating direction is to increase the weight on socially rational update, its weighting factor $w_{i,sl}$ will be increased by $\delta$, and the value of $w_{i,il}$ will be decreased by $\delta$ accordingly (Line 11–17). During each period, each agent also keeps record of its performance, i.e., the average payoff obtained within the period. The updating direction is adjusted by comparing the relative performance between two consecutive periods. Each agent can switch its current updating direction to the opposite one if its performance in the current round is worse than that of previous period (Line 5–10). For the first adjustment period, since there is no previous performance record for comparison, each agent simply keeps its initial updating direction unchanged and updates the weighting factors accordingly (Line 19–26).

If an agent $i$ updates its $Q$-values in the individually rational manner, it acts as an individual rational entity which uses the information of possible individual payoffs received by performing different actions. Let us denote $S^t$ as the set of information consisting of the action choices and payoffs of itself and the agents of the same role from the other $M$ groups, i.e., $S^t_{il} = \{\langle a^t_i, r^t\rangle, \langle b^t_1, r^t_1\rangle, \langle b^t_2, r^t_2\rangle, \ldots, \langle b^t_M, r^t_M\rangle\}$. Here $\langle a^t_i, r^t\rangle$ is the action and payoff of agent $i$ itself, and the rest are the actions and payoffs of other $M$ agents with the same role from other $M$ groups. At the end of each round $t$, agent $i$ picks action $a$ with the highest payoff (randomly choosing one action in case of a tie) from the set $S^t_{il}$ and updates its $Q$-value of action $a$ according to Eq. 4.4 by assigning $r_{i,il}$ to the value of $r(a) * \text{freq}(a)$. Here $r(a)$ is the highest payoff of action $a$ among all elements in the set $S^t_{il}$, and freq$(a)$ is the frequency that action $a$ occurs with the highest reward $r(a)$ among the set $S^t_{il}$.

If an agent $i$ learns its $Q$-values based on socially rational update, it updates its $Q$-values based on the joint benefits of agents in the same group. Let us denote $S^t$ as the set of information consisting of the action choices and group payoffs of itself and the agents of the same role from the other $M$ groups, i.e., $S^t_{sl} = \{\langle a^t_i, R^t\rangle, \langle b^t_1, R^t_1\rangle, \langle b^t_2, R^t_2\rangle, \ldots, \langle b^t_M, R^t_M\rangle\}$. Here $\langle a^t_i, R^t\rangle$ is the action and sum of the payoffs of agent $i$ and its interaction partner, and the rest are the actions and group payoffs of the agents with the same role from other $M$ groups. Similarly to the case of individually rational update, at the end of each round $t$, each agent $i$ picks the action $a$ with the highest group payoff (randomly choosing one action in case of a tie) from the set $S^t_{sl}$ and updates the value of $Q(s, a)$ according to Eq. 4.5 by assigning $r_{i,sl}$ to the value of $R(a) * \text{freq}(a)$. Here $R(a)$ is the highest group payoff of action $a$ among all elements in the set $S^t_{sl}$, and freq$(a)$ is the frequency that action $a$ occurs with the highest group reward $R(a)$ among the set $S^t_{sl}$.

### 4.2.2 Analysis of the Learning Performance Under the Social Learning Framework

First, let us suppose that the agents can finally learn a consistent policy of coordinating on one socially optimal outcome $(x_s, y_s)$, i.e., each agent always chooses action $x_s$ as the row player and action $y_s$ as the column player. Since each agent is assigned as the row or column player randomly each time, the average payoff of each agent approaches $\bar{p} = \frac{p_{x_s}+p_{y_s}}{2}$ in the limit. This indicates that the agents are able to achieve both fair and highest payoff whenever they can learn a consistent policy of coordinating on one socially optimal outcome. Accordingly, we can have the following theorem.

**Theorem 4.1** *If the agents can finally learn a consistent policy of coordinating on one socially optimal outcome $(x_s, y_s)$, then each agent will obtain both equal and highest payoff of $\bar{p}$ on average; if every agent receives an average payoff of $\bar{p}$, then they must have learned to coordinate on some socially optimal outcome(s).*

*Proof* It is obvious that the agents' average payoffs are fair (equal) from the above analysis. Let us suppose that there exists an agent $i$ which can achieve an average payoff of $\bar{p}' > \frac{p_{x_s}+p_{y_s}}{2}$ by coordinating on nonsocially optimal outcomes. Since the agents within the system are symmetric, all other agents in the system should receive the same average payoff of $\bar{p}'$ as well. This contradicts with the definition of socially optimal outcome, and thus no agent can ever achieve an average payoff higher than $\frac{p_{x_s}+p_{y_s}}{2}$. By applying similar analysis, we can know that the achievement of an average payoff of $\frac{p_{x_s}+p_{y_s}}{2}$ also must be done through coordinating on socially optimal outcome(s). □

When all agents behave as socially rational entities to perform update on their $Q$-functions (in the first period), it is not difficult to see that for any general-sum game $G$, the agents are actually learning over its corresponding fully cooperative game $G'$ where the payoff profile $(p_x, p_y)$ of each outcome $(x, y)$ in $G$ is changed to $(p_x + p_y, p_x + p_y)$ in $G'$. From the definition of socially rational outcome, we know that every socially rational outcome in the original game $G$ corresponds to a Pareto-optimal Nash equilibrium in the transformed game $G'$ and vice versa. Therefore, the goal of learning to achieve socially rational outcomes in game $G$ becomes equivalent to learning to achieve Pareto-optimal Nash equilibrium in the transformed cooperative game $G'$. Based on Theorem 4.1, we know that the socially rational agents can obtain both fair and highest payoff as long as they are able to learn to converge to a Pareto-optimal Nash equilibrium in $G'$. The problem of learning toward Pareto-optimal Nash equilibrium in cooperative games under the social learning framework has been investigated in previous work [2], and their empirical results show that the agents can always successfully coordinate on some Pareto-optimal Nash equilibrium even in the challenging cooperative games: the climbing game and the penalty game [9].

Based on the empirical results in [2] mentioned above, now let us analyze another situation when all agents behave as individually rational entities to update their $Q$-functions (in the second period). From Theorem 4.1, we know that the each agent's average payoff $\overline{p}$ in this situation must be no better than that when all agents act in the socially rational manner, i.e., $\overline{p} \leq \frac{p_{x_S}+p_{y_S}}{2}$. We discuss in the following two cases.

**Case 1:** $\overline{p} < \frac{p_{x_S}+p_{y_S}}{2}$ In this case, the agents will change their $Q$-function update schemes back to the socially rational update starting from the third period. As previously mentioned, previous work [2] indicates that the agents can converge to some socially optimal outcome under the social learning update scheme,[2]; thus, the agents' average performance can be stabilized at the highest level $\frac{p_{x_S}+p_{y_S}}{2}$ at the end of the third period. According to Algorithm 3, the agents will continue behaving as socially rational entities in the following periods, and thus it is expected that some socially optimal outcome will be converged to forever.

**Case 2:** $\overline{p} = \frac{p_{x_S}+p_{y_S}}{2}$ In this case, from Theorem 4.1, we know that the agents must have somehow successfully coordinated on socially optimal outcome(s) by behaving in individually rational manner. As long as the agents' average performance is not decreased, the agents will continue acting as individually rational entities according to Algorithm 3, and socially optimal outcome(s) will be always achieved. In case their average performance is decreased due to mis-coordination on some nonsocially optimal outcomes in later period, i.e., $\overline{p} < \frac{p_{x_S}+p_{y_S}}{2}$, it becomes the same situation with case 1, and the same analysis as case 1 can be applied here.

Overall, from previous analysis, we can see that it is expected that the agents can adaptively learn to coordinate on some socially optimal outcome and thus receive both fair and highest payoff on average in any general-sum games. In next section, we are going to empirically evaluate the performance of the social learning framework through extensive simulations.

### *4.2.3 Experimental Evaluations*

In this section, we empirically evaluate the performance of the social learning framework and investigate the influences of different factors on the learning performance. First, we use two well-known examples (the prisoner's dilemma game and learning "rules of the road") to illustrate the learning dynamics and compare the performance of the learning framework with that of previous work in Sects. 4.2.3.1 and 4.2.3.2, respectively.[3] In Sect. 4.2.3.3, we further evaluate the performance of

---

[2]Notice that socially optimal outcomes here correspond to Pareto-optimal Nash equilibria in the transformed game $G'$ under the social learning update scheme.

[3]Note that these two examples are the most commonly adopted test beds in previous work [4, 5, 20, 22].

the social learning framework under a larger set of general-sum games to illustrate the generality of our social learning framework. Finally, to evaluate the robustness and sensitivity of the social learning framework, the influences of different factors (against purely selfish agents, information sharing degree, and fixed policy agents) on the learning performance are investigated from Sects. 4.2.3.4, 4.2.3.5 and 4.2.3.6.

#### 4.2.3.1 Examples

Prisoner's Dilemma Game

Prisoner's dilemma game is one of the most well-known games which have attracted much attention in both game theory and multiagent system literatures. One typical example of the prisoner's dilemma game is shown in Fig. 4.9. For any individually rational player, choosing defection (action $D$) is always its dominating strategy; however, mutual defection will lead the agents to achieve the inefficient outcome $(D, D)$, while there exists another socially optimal outcome $(C, C)$ in which each agent could obtain a much higher and fair payoff. Figure 4.10 shows the average frequency of each joint action that the agents learn to coordinate on under the social learning framework in a population of 100 agents. We can see that the agents can eventually learn to converge to the outcome $(C, C)$, which corresponds to the policy of "choosing action C as both row and column players." During the first period, all agents update their policies based on socially rational update scheme, and the policy of "always choosing action $C$" is reached after approximately 250 rounds. It is worth noting that the outcome $(C, C)$ actually corresponds to the only Nash equilibrium in the cooperative game after transformation. Starting from the second period, the agents change their update scheme to individually rational update, and they learn to converge to the policy of "always choosing action $D$" and thus converging to outcome $(D, D)$ starting from the middle point of the second period. Since the average payoff obtained from mutual cooperation in the first period is much higher than that obtained from mutual defection in the second period, the agents changed their update scheme back to socially rational update at the beginning of the third period and will stick to this scheme forever. Therefore, mutual cooperation $(C, C)$ is always achieved among the agents starting from the middle point of the third period.

**Fig. 4.9** An instance of prisoner's dilemma game

| 1's payoff, 2's payoff | | Player 2's action | |
|---|---|---|---|
| | | C | D |
| Player 1's action | C | 3, 3 | 0, 5 |
| | D | 5, 0 | 1, 1 |

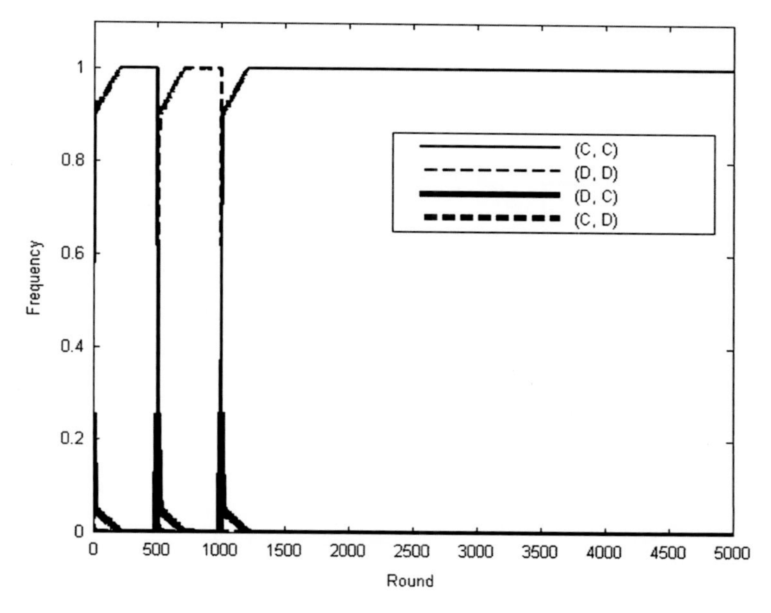


**Fig. 4.10** Frequency of reaching different joint actions in prisoner's dilemma game averaged over 100 times (the length of each period $m = 500$ and adjustment degree $\delta = 1.0$)

### Learning "Rules of the Road"

The problem of learning "rules of the road" [18] can be used to model the practical car-driving coordination scenario: who yields if two drivers arrive at an intersection at the same time from neighborhood roads. Both drivers would like to go across the intersection first without waiting; however, if both drivers choose to not yield to the other side, they will collide with each other. If both of them choose to wait, then neither of them can go across the intersection, and it is a waste of their time. Their behaviors are well coordinated if and only if one of them chooses to wait and the other driver chooses to go first. This problem can be naturally modeled as a two-player two-action anti-coordination game (see Fig. 4.7 where action $C$ corresponds to "Yield" and action $D$ corresponds to "Go"). The natural solution of this problem is that the agents adopt the policy of "yields to the driver on left (right)," and this solution is optimal in that both agents can obtain equal and highest payoff on average provided that their chances of driving on the left or right road are the same. Figure 4.11 shows the average frequency of different joint actions that the agents learn to converge to within a population of 100 agents. The results are obtained averaging over 100 independent runs, in which half of the runs the agents learn to acquire the policy of "yielding to the driver on the left" $((D, C))$ and acquire the policy of "yielding to the driver on the right" $((C, D))$ the rest of the times. Another observation is that the agents learn to converge to one of the previous two optimal policies under individual learning update schemes (transition

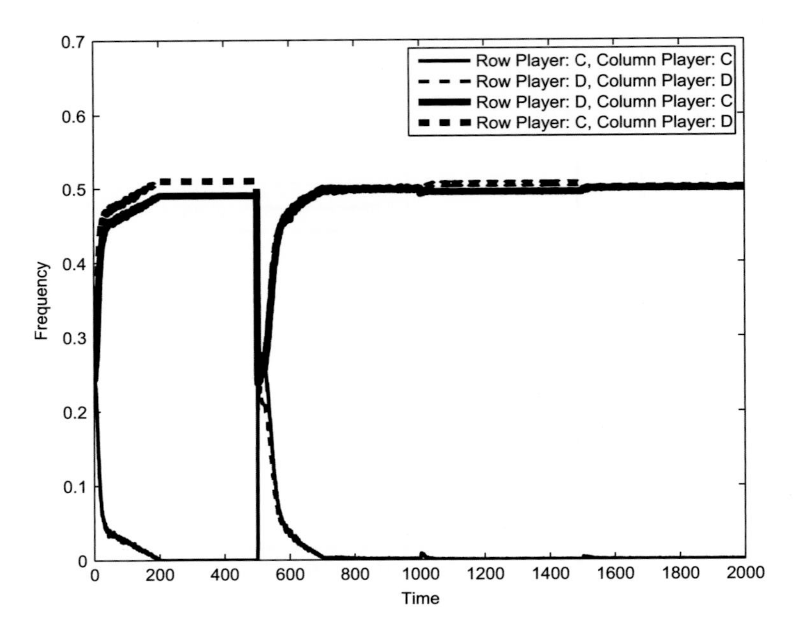


**Fig. 4.11** Frequency of reaching different joint actions in anti-coordination game averaged over 100 times (the length of each period $m = 500$ and adjustment degree $\delta = 1.0$)

| A's payoff, B's payoff | | Agent B's action | |
|---|---|---|---|
| | | C | D |
| Agent A's action | C | 0, 0 | 5, 4 |
| | D | 2, 15 | 0, 0 |

**Fig. 4.12** Asymmetric anti-coordination game: the agents need to coordinate on the outcome $(D, C)$ to achieve socially rational solution

from social learning update to individual learning update at 500th round). This is in accordance with the results found in previous work [18], in which the authors investigate whether the agents can learn to coordinate on one of the previous two optimal policies if they learn based on their private information only (equivalent with the individually rational update here).

Next we consider an asymmetric version of the anti-coordination game shown in Fig. 4.12. Different from the symmetric anti-coordination game in which it is sufficient for the agents to coordinate on the complementary actions (either $(C, D)$

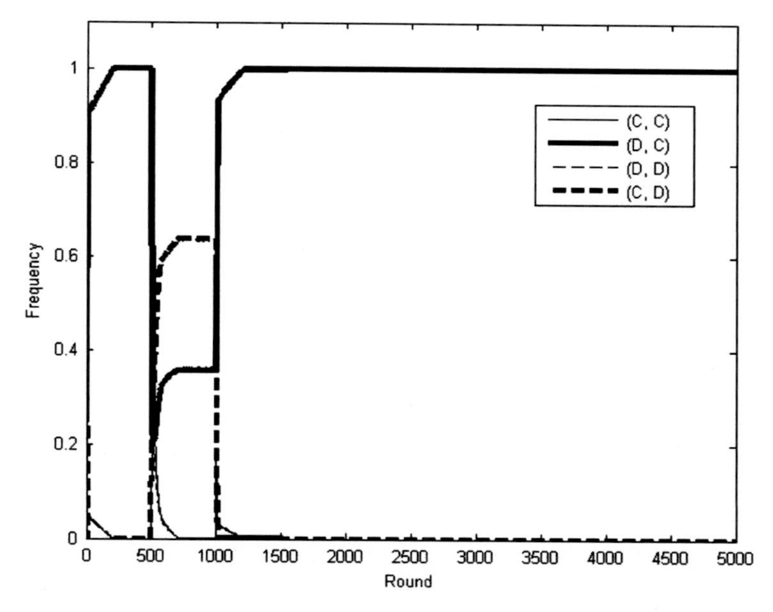


**Fig. 4.13** Average frequency of achieving socially optimal outcome $(D, C)$ under the learning framework in asymmetric anti-coordination game (averaged over 100 runs)

or $(D, C)$), here the row and column agents need to coordinate their joint actions on the unique outcome $(D, C)$ only in order to achieve the socially optimal solution. Specifically, each agent needs to learn the unique policy of $row \rightarrow D, column \rightarrow C$, i.e., selecting action $D$ as the row player and selecting action $C$ as the column player. Figure 4.13 shows the frequency of each outcome the agents achieve averaged over 100 runs under the social learning framework. We can observe that the agents learn to converge to the socially rational outcome $(C, D)$ during the first period under socially rational update scheme, and each agent receives an average payoff of 8.5 after convergence. Starting from the second period, the agents begin to adopt individually rational update for the purpose of exploration, and either $(C, D)$ or $(D, C)$ is converged to with probability of approximately 0.6 and 0.4, respectively. If the agents learn to converge to $(C, D)$ during the second period, then the average payoff each agent receives after convergence is only 4.5. Accordingly, the agents will have the incentive to change their update scheme back to socially rational update, and they will stick to this scheme thereafter since they can always receive the higher average payoff of 8.5. Notice that different from the symmetric anti-coordination game, only using individually rational update is not sufficient to guarantee the convergence to socially optimal outcome $(D, C)$ in asymmetric case. Under individually rational update scheme, $(C, D)$ has much higher opportunities to be converged to even though $(D, C)$ is the only socially optimal outcome. Therefore, socially rational update is necessary to guarantee that socially optimal outcome $(D, C)$ can be achieved in this example.

From previous experimental results in Sects. 4.2.3.1 and 4.2.3.1, we have shown that different update schemes (social learning update or individual learning update) may be required to achieve socially optimal solutions in different interaction scenarios, and also the agents using our adaptive update strategy are able to successfully learn to choose the most suitable update scheme accordingly. Next we are going to evaluate the learning performance of the social learning framework by comparing it with previous work using the previous examples.

#### 4.2.3.2 Comparison with Previous Work

In this section, we compare the performance under the social learning framework with two learning frameworks proposed in previous work [4, 5]. Since all previous work is only applicable in symmetric games, here we evaluate their performance under two types of representative games: the prisoner's dilemma game (Fig. 4.9) and anti-coordination game (Fig. 4.7), which are the most commonly adopted test beds in previous work [4, 5, 25]. Figures 4.14 and 4.15 show the dynamics of the percentage of agents coordinating on socially optimal outcomes as a function of rounds for the learning frameworks in both [5] and [4] under the prisoner's dilemma game and anti-coordination game, respectively. We can see that both previous learning frameworks can lead high percentage of agents to achieve coordination on socially optimal outcomes on average; however, there is always still a small proportion of agents that fails to do so all the time. This phenomenon is particularly obvious for the paired reproduction-based learning framework in [4], in which the deviation of the percentage of agents coordinating socially optimal outcomes is very high since the agents under their learning framework use evolutionary learning (copy and mutation) to make their decisions. In their approach, mutation plays an important role in evolving high level of coordinations on socially optimal outcome,

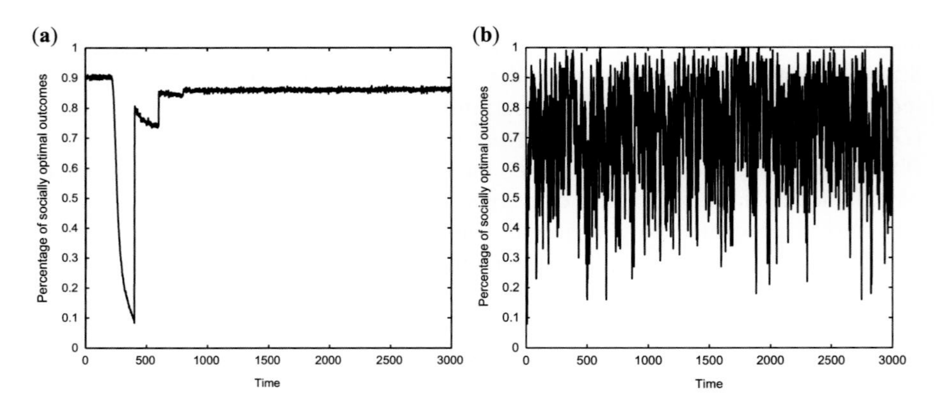


**Fig. 4.14** Learning performance of previous work in prisoner's dilemma game. (**a**) Results using the learning framework in [5]. (**b**) Results using the learning framework in [4]

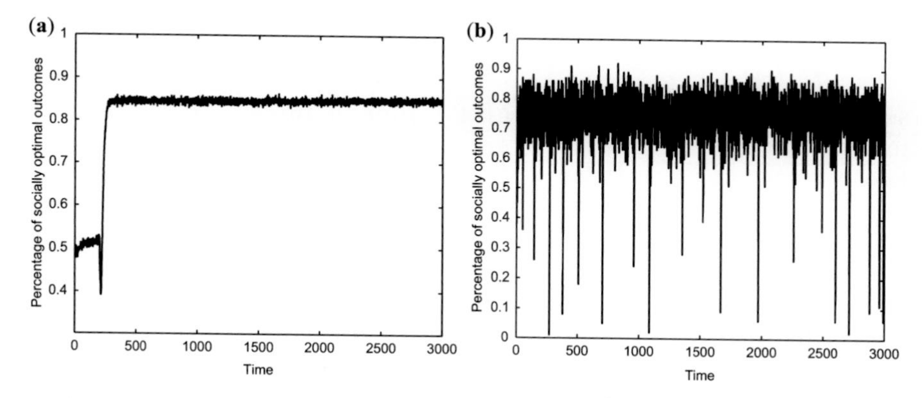


**Fig. 4.15** Learning performance of previous work in anti-coordination game. (**a**) Results using the learning framework in [5]. (**b**) Results using the learning framework in [4]

while the corresponding side effect is that the deviation of the percentage of socially optimal outcome is very high. In contrast, under our socially learning framework, all the agents can stably learn to coordinate on socially optimal outcomes in both the prisoner's dilemma game (see Fig. 4.10) and anti-coordination game (see Fig. 4.11).

#### 4.2.3.3 General Performance in Two-Player General-Sum Games

In this section, we further evaluate the performance of the social learning framework in a large set of two-player general-sum games. All games are generated randomly with the payoffs between the range of 0 and 100. Table 4.1 lists the average success rate in terms of the percentage of games in which full coordination on socially optimal outcomes can be achieved when the values of adjustment period length $m$ and the acceptance threshold *thres* vary. All the results are averaged over 100 randomly generated games. First, we can see that under our social learning framework the agents can achieve a high success rate with the highest value of 0.97 when $m = 2000$ and $thres = 0$. However, there still exist 3 % of games in which the agents cannot achieve full coordination on socially optimal outcomes. The underlying reason is that in those games there exists a nonsocially optimal outcome such that the agents' average payoff by coordinating on this outcome is very close to that by coordinating on the socially optimal outcome, and thus some agents may fail to distinguish which one is better, and finally learn the incorrect update scheme and fail to learn their individual actions corresponding to the socially optimal outcome, which thus results in certain degrees of mis-coordinations on nonsocially optimal outcomes.

Next we analyze the influence of the values of *thres* and $m$ on the success rate of coordinating on socially optimal outcomes. The value of the threshold *thres* reflects the tolerance degree of the performance of the current learning period. The larger

**Table 4.1** Success rate in terms of percentage of games in which full coordination on socially optimal outcomes can be achieved in randomly generated general-sum games for different values of $m$ and *Threshold*

| Success rate ($m$/*Threshold*) | 0 | 0.001 | 0.01 | 0.05 | 0.1 | 0.2 |
|---|---|---|---|---|---|---|
| 1000 | 0.586 | 0.624 | 0.79 | **0.916** | 0.812 | 0.676 |
| 1500 | 0.504 | 0.554 | 0.856 | **0.92** | 0.84 | 0.69 |
| 2000 | **0.97** | 0.968 | 0.964 | 0.962 | 0.806 | 0.738 |

the length of each adjustment period ($m$) is, the more accurate estimation of the actual performance of each period we can obtain. When the estimation of the actual performance of each period is not accurate enough (the value of $m$ is relatively small), setting the value of the threshold *thres* too large or small can result in high rate of mis-coordination on nonsocially optimal outcomes. If the value of *thres* is too small, it is likely that some agents may choose the wrong update scheme as their next-round update schemes due to the noise of inaccurate estimations of the performance in each round; if the value of *thres* is too large, it is obvious that some agents may make wrong judgments on the optimal update scheme due to high tolerance degree, thus resulting in certain degrees of mis-coordination. Indeed, this can be verified from Table 4.1 that when the length of adjustment period is small ($m$ = 1000 or 1500), the highest success rate of achieving full coordination on socially optimal outcomes is achieved when the value of the threshold is set to 0.05. Either decreasing or increasing the value of the threshold *thres* will decrease the success rate of coordinating on socially optimal outcomes. When the length of adjustment period is further increased to 2000, the agents are able to get more accurate estimation of the performance of each period; thus, the highest success rate is achieved when the threshold is set to the minimum value of 0.

#### 4.2.3.4 Against Purely Individually Rational Agents

The successful coordination on socially optimal outcomes in the social learning framework requires the cooperation of all agents in the system, i.e., each agent should have the intention to follow the learning strategy proposed in Sect. 4.2.1.3. However, in an open multiagent environment, it is possible that there exist some selfish agents that would like to use this opportunity to exploit those agents that update their $Q$-values in the socially rational manner to the benefits of their own. For example, consider the prisoner's dilemma game, to achieve socially optimal outcome $(C, C)$, it requires all agents cooperatively to learn to adopt socially rational update scheme and finally learn to choose action $C$ each round. However, it is obvious that they can be easily exploited by the small amount of purely individually rational agents, which can learn to choose action $D$ and thus obtain a much higher payoff. In this section, we empirically show that the agents following the strategy in Sect. 4.2.1.3 can automatically learn to adopt the individually rational update scheme back when the number of selfish agents is too much such that there are

**Table 4.2** Average payoffs of agents following the strategy in Sect. 4.2.1.3 and SAs in the prisoner's dilemma game with the percentage of SAs varies

| Average payoffs | 10 % | 20 % | 30 % | 40 % | 50 % | 60 % | ≥70 % |
|---|---|---|---|---|---|---|---|
| Agents following strategy in Sect. 4.1.1.2 | 2.75 | 2.45 | 2.2 | 1.81 | 1.48 | 1.08 | 1 |
| Selfish agents (SAs) | 4.39 | 4.15 | 3.8 | 3.43 | 3.02 | 1.28 | 1 |

no benefits for them to adopt socially rational update scheme any more. In this way, the agents can effectively prevent from being exploited by those selfish agents.

Table 4.2 shows the average payoffs of the agents following the strategy in Sect. 4.2.1.3 and purely selfish agents when the percentage of purely selfish agents varies in the prisoner's dilemma game. If all agents are purely selfish (corresponding to the case of 100 % of purely selfish agents), then all agents will learn to converge to choose action $D$ and obtain an average payoff of 1. However, when there is only a relatively small percentage of selfish agents, the rest of agents can still obtain an average payoff higher than 1 by performing socially rational update. This corresponds to the cases when the percentage of purely selfish agents are less than 70 % in Table 4.2, in which the rest of agents still learn to adopt socially rational update and finally choose action $C$ and thus obtain an average payoff higher than 1.0. When the percentage of purely selfish agents becomes larger than 70 %, then they can realize that there is no benefit for them to adopt socially rational update any more and thus switch to individually rational update and finally learn to choose action $D$, and all agents obtain an average payoff of 1.

#### 4.2.3.5 Influences of the Amount of Information Shared Among Agents

One major difference between the social learning framework and the commonly adopted setting of two-agent repeated interactions is that the agents have more information (other groups of interaction agents in the population) at their disposal when they learn their coordination policies. Therefore, it is particularly important to have a clear understanding of the influence of the amount of information shared among the agents (the value of $M$) on the system-level's performance. To this end, in this section, we evaluate the influence of the value of $M$ on the learning performance of the system using the symmetric anti-coordination game in Fig. 4.7. Figure 4.16 shows the dynamics of the average payoff of each agent as a function of rounds with different values of $M$ under the socially rational update scheme. We can see that each agent's average payoff increases more quickly with the increase of the value of $M$. Intuitively, when the value of $M$ becomes larger, it means the agents are exposed to more information each round and it is expected that they are able to learn their policies in a more informed way and thus coordinate their actions on socially optimal outcomes more efficiently. However, it is worth noticing that the time required before reaching the maximum average payoff is approximately the same for different values of $M$.

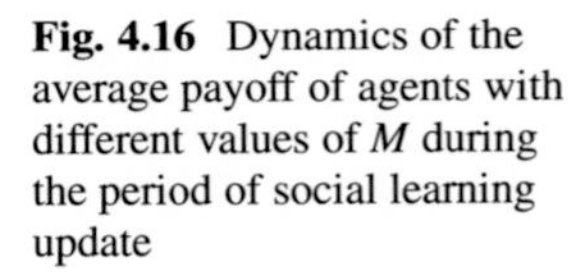

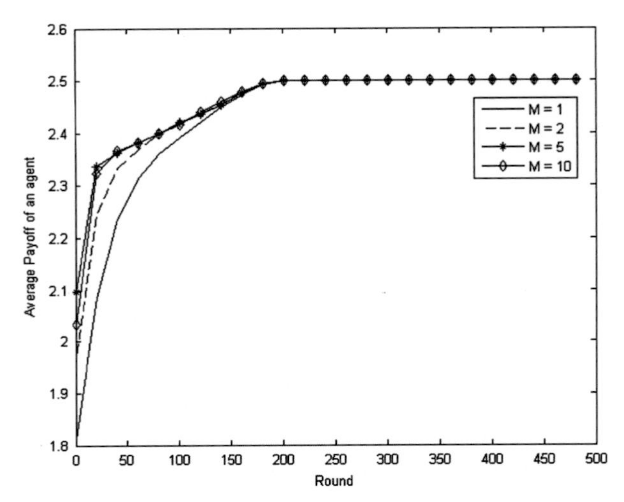


**Fig. 4.16** Dynamics of the average payoff of agents with different values of $M$ during the period of social learning update

#### 4.2.3.6 Influence of Fixed Agents

We have shown that both socially optimal outcomes $(C, D)$ and $(D, C)$ have equal chances to be converged to in the symmetric anti-coordination game in Sect. 4.2.3.1. This is reasonable and expected since the game itself is symmetric and all agents are identical; thus, there is no bias toward any of the two outcomes. We also have investigated the influence of game structure on the convergence results of the system when the game itself is not symmetric. In this section, we consider another situation when there exist some agents that adopt a fixed policy without any learning ability and evaluate how the existence of small number of fixed agents can influence the convergence outcome of the system.

Figure 4.17 shows the frequency of each socially optimal outcome can be converged to when different numbers of fixed agents are inserted into the system using the symmetric anti-coordination game (Fig. 4.7). We assume that the fixed agents always make their decision following the policy of "choosing action $D$ as the row player and choosing action $C$ as the column player." When there is no fixed agents inserted into the system, both socially optimal outcomes $(C, D)$ and $(D, C)$ can be converged to with equal probability. When we insert only one fixed agent into the system, it is surprising to find that the probability of converging to the outcome $(D, C)$ (corresponding to converging to the policy of the fixed agent) is increased to 0.7! When the number of fixed agents is further increased, the probability of converging to $(D, C)$ is increased as well. With only four fixed agents inserted into the system, the agents converge to the outcome $(D, C)$ with approximately 100 % chance. There might thus be some truth to the adage that most fashion trends are decided by a handful of trendsetters in Paris! [18].

Figure 4.18 illustrates the detailed dynamics of the frequency changes of different outcomes in a population of 100 agents when only one fixed agent is inserted into

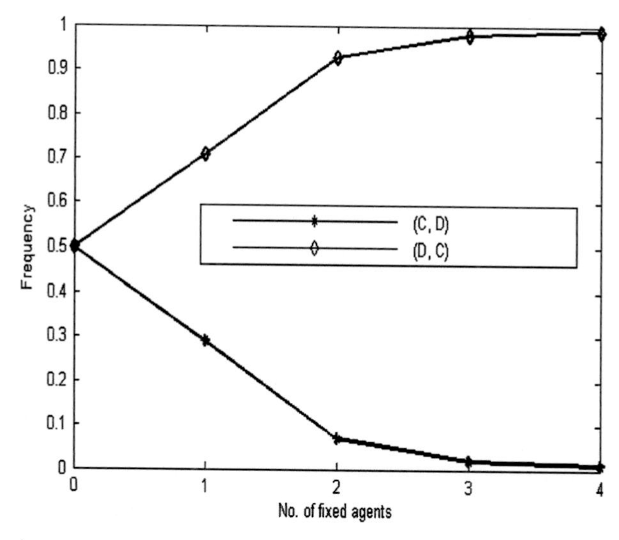


**Fig. 4.17** Average frequency of socially optimal outcomes converged to under the learning framework in asymmetric anti-coordination game with different number of fixed agents (averaged over 100 runs)

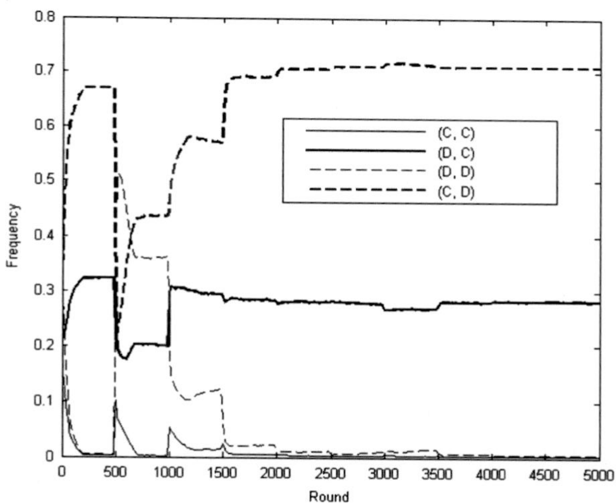


**Fig. 4.18** Average frequency of achieving each outcome under the learning framework in asymmetric anti-coordination game with one fixed agent (averaged over 100 runs)

the system. First, we can see that the agent with fixed policy has significant influence on the convergence outcomes in the first period under socially rational update, i.e., about 70 % chances of converging to $(C, D)$ instead of 50 % without any fixed agent. During the second period under individually rational update, there is about 37 % chances that the agents mis-coordinate their action choices (reaching $(D, D)$) due to the existence of the fixed policy agent, and also the chance of converging to $(C, D)$ is higher than that of converging to $(D, C)$. The agents change their update scheme back to socially rational update in the third period, and the chance of converging to $(C, D)$ is gradually increased to the maximum value (approximately 70 % of the runs) by the end of the fourth period.

## 4.3 Achieving Socially Optimal Allocations Through Negotiation

Previous two sections focus on the general problems of multiagent coordination toward socially optimal solutions, in which the interactions among each pair of agents are modeled as different types of normal-form games (cooperative games and general-sum games). In this section, we turn to investigate a concrete and important multiagent interaction problem—multiagent resource allocation problem—and propose a negotiation framework for agents to achieve socially optimal solutions, i.e., socially optimal allocation of resources. The problem of resource allocation among multiple agents through negotiation has received much attention in multiagent system research [26–29]. Multiagent negotiation over resources can be applied in a variety of practical domains such as network bandwidth allocation, robotics, and grid computing [30, 31]. On one hand, a rational agent prefers to increase its own utility as much as possible through negotiation; on the other hand, the desirable goal, from the system's perspective, is to allocate the resources among agents as efficient as possible.

When the agents negotiate over a bundle of resources, the negotiation process can be quite complex due to the huge space of all possible deals. The complexity of negotiation with respect to different goals and different kinds of deals and utility functions allowed has been theoretically investigated in previous work [28, 32, 33]. A natural direction to tackle the complexity of negotiation is to design efficient protocols to guide the negotiation process and efficient strategies for agents to improve their negotiation ability. Following this direction, Chevaleyre et al. [27] propose protocols to guide the agents to reach optimal allocations by exploiting the structural properties of agents' utility functions used for modeling their preferences over resources. However, their protocols are only applicable to the specific case that the agents' utility functions are $k$-additive and in tree structured form and also they assume that each agent's utility function is common knowledge to the system. There also exist other protocols putting similar assumptions on the agents' utility functions [34, 35]. Such kind of protocols is of limited usage due to the constraints on the agents' utility functions, and they are not applicable in the scenarios in which the agents are not willing to reveal this private information.

To address this issue, Saha and Sen [29] propose a negotiation protocol PONOMIR by assuming that the agents' preferences are private information. Their protocol is guaranteed to converge to Pareto-optimal allocation. However, under Pareto-optimal allocation, it is possible that the resources may not be fully utilized by the agents, which thus reduces the system's efficiency as a whole. For example, in the context of e-commerce application, we usually aim to maximize the average profit generated by the negotiating agents [26]. A better criterion for this kind of scenarios is the notion of socially optimal allocation [36], i.e., the sum of agents' utilities over the resources that they are assigned under the allocation is maximized. Socially optimal allocation can be considered as a special case of Pareto-optimal allocation under which the sum of the agents' utilities is maximized. As pointed out in [32], to guarantee that socially optimal allocation can always be reached, side payments among agents have to be allowed to compensate those agents suffering from loss during negotiation. To this end, we propose a negotiation framework for agents which is guaranteed to achieve socially optimal allocations.

### 4.3.1 Multiagent Resource Allocation Problem Through Negotiation

The specific negotiation framework we consider for multiple rational agents to negotiate over a number of indivisible resources is described as follows. There exist a set $\mathcal{R} = \{R_1, R_2, \ldots, R_N\}$ of $N$ indivisible resources and a set $\mathcal{A} = \{1, 2, \ldots, n\}$ of $n$ agents in the system. Each agent $i$ has its own preference over different bundles of resources, which is represented by the utility function $u_i : 2^{\mathcal{R}} \rightarrow \mathbb{Z}$, where $\mathbb{Z}$ is the set of integers. We assume that the agents' utility functions are monotonic, i.e., assigning an additional resource to an agent will not decrease its utility. Each agent's preference over the resources is its private information and is inaccessible by other agents.

Next we give the definitions of useful concepts for the convenience of description. An allocation $A$ refers to a particular partition of the set $\mathcal{R}$ of resources among the set $\mathcal{A}$ of agents. The resources assigned to agent $i$ under an allocation $A$ is denoted as $A(i)$. Give an allocation $A_1$ over the set $\mathcal{R}$ of resources, the agents can propose a deal to exchange their resources to improve their own utilities. This process will take the agents from the original allocation $A_1$ to another one $A_2$, and the deal can be represented as a pair of allocations $(A_1, A_2)$ [32].

**Definition 4.1** A *deal* is a pair $\delta = (A_1, A_2)$ where $A_1$ and $A_2$ are allocations of the resources over $\mathcal{R}$ with $A_1 \neq A_2$.

When the agents negotiate with one another, we assume that they can use money to compensate the agent who suffers from utility loss during negotiation. Here we use a payment function to indicate how much money each agent receives or gives away when a deal is reached.

**Definition 4.2** A payment function is a function $p : \mathcal{A} \rightarrow \mathbb{Z}$ which satisfies the condition that $\sum_{i \in \mathcal{A}} p(i) = 0$.

Here if $p(i)$ is positive, $|p(i)|$ means the amount of money that agent $i$ receives in the deal; if $p(i)$ is negative, $|p(i)|$ indicates the amount of money that agent $i$ has to give away in the deal.

Note that a deal $(A_1, A_2)$ can be implemented if and only if it can be accepted by all agents in the system. In other words, each agent's total utility under the new allocation $A_2$ must be no less than its total utility under allocation $A_1$. In this situation, the deal $(A_1, A_2)$ is considered as an individually rational deal.

Alternatively, we can say that allocation $A_2$ is an individually rational allocation for all agents with respect to $A_1$.

**Definition 4.3** A deal $(A_1, A_2)$ is individually rational iff there exists a payment function $p$ such that the following conditions are satisfied: $\forall i \in \mathcal{A}, u_i(A_2(i)) + p(i) \geq u_i(A_1(i))$.

We assume that the agents are *altruistic-individually rational*, which satisfies the following properties: (1) each agent is individually rational in that it only accepts the deals that will not decrease its individual utility; (2) each agent is altruistic in that it only requires to increase its own utility by the minimum unit of the utility,[4] whenever it proposes a deal to others. In other words, the agents are assumed to be selfish but not extremely greedy.

Our goal is to achieve a socially optimal allocation $A^*$ through distributed negotiation among agents, i.e., the *utilitarian social welfare* of the system $\sum_{i \in \mathcal{A}} u_i(A^*(i))$ [37] is maximized. Note that the agents' payments are not taken into account here since the sum of all agents' payments equals to zero. In next section we will present the protocol APSOPA, which leads the agents to reach socially optimal allocation.

### 4.3.2 The APSOPA Protocol to Reach Socially Optimal Allocation

For the ease of description, we present the APSOPA protocol in the bilateral negotiation scenario. The negotiation protocol consists of three stages. In the first stage, the agents are allowed to allocate the resources in the way specified by an allocation procedure, and an initial allocation $A_0$ is obtained using strict alternation protocol [38]. Based on the initial allocation $A_0$, in the second stage, the agents begin the distributed negotiation process over individual resources. In the last stage, the agents will make further negotiations over the candidate allocations obtained in previous stages. The resources will be allocated according to the final allocation reached in this stage. We first define some concepts useful for describing our protocol in Sect. 4.3.2.1 and then present our protocol in Sect. 4.3.2.2.

#### 4.3.2.1 Concept Definition

There are a few concepts that are useful in the description of the protocol.

[4]Note that the minimum unit of the agents' utilities is 1 here, since the utility function is defined in integers.

**Negotiation Tree** Consider the negotiation scenario involving a set $\mathcal{R} = \{R_1, R_2, \ldots, R_N\}$ of resources and a set $\mathcal{A} = \{1, 2\}$ of two agents. The corresponding negotiation tree is a binary tree with maximum height of $N$, i.e., the number of resources. We define that the level of the root node is 0, and thus the maximum possible level of the negotiation tree is $N$. The root node of the negotiation tree represents the initial situation when none of the resources has been allocated yet. For each node in level $l-1(1 \leq l \leq N)$ of the negotiation tree, its left branch represents the case when the resource $R_l$ is allocated to agent 1, and its right branch represents the complementary case that $R_l$ is allocated to agent 2. Thus each non-leaf node in level $l$ of the negotiation tree corresponds to a partial allocation over the first $l$ resources with the rest of $N-l$ resources unallocated. Each leaf node at level $N$ in the negotiation tree corresponds to one complete allocation over the set $\mathcal{R}$ of resources.

**Best Possible Utility** For each node $n$ in the negotiation tree, each agent $i$ has its best possible utility $BPU_i(n)$ over the resources, which is calculated as the maximum utility this agent can possibly receive based on the partial allocation of this node. Let us assume that the set of resources allocated to agent 1 following the partial allocation specified by node $n$ is $\mathcal{P}$, and the corresponding set of resources allocated to agent 2 is $\mathcal{Q}$, thus the set of remaining resources is $\mathcal{L} = \mathcal{R} \setminus \mathcal{P} \setminus \mathcal{Q}$. For those unallocated resources $\mathcal{L}$, the best case for each agent would be that all of them are assigned to the agent itself. Thus the best possible utilities for the agents on node $n$ are $BPU_1(n) = u_1(\mathcal{P} \cup \mathcal{L})$ and $BPU_2(n) = u_2(\mathcal{Q} \cup \mathcal{L})$, respectively. If node $n$ is a leaf node at level $N$, then agent $i$'s best possible utility $BPU_i(n)$ is simply equal to its actual utility over the corresponding allocation specified by node $n$.

**Payment Information** Since side payment between agents is allowed, the agents can also utilize money to make compensation to the agent suffering from utility loss during negotiation. Each node $n$ of the negotiation tree is associated with a pair of payment information $\langle P_n(1), P_n(2) \rangle$. If the value of $P_n(i)$ ($\forall i \in \{1, 2\}$) is positive, it represents the maximum amount of money that agent $i$ is willing to pay for reaching the partial allocation specified by node $n$. Similarly, if the value of $P_n(i)$ ($\forall i \in \{1, 2\}$) is negative, it represents the minimum amount of money that agent $i$ must receive to have the incentive to accept the partial allocation specified by node $n$. Since the agents have reached an initial allocation $A_0$ before starting the negotiation in Stage 2, the values of the payoff information of both agents for node $n$ are the differences between their best possible utilities over the current node and their utilities over $A_0$. That is, $P_n(1) = BPU_1(n) - u_1(A_0(1))$ and $P_n(2) = BPU_2(n) - u_2(A_0(2))$.

To better understand the above concepts, consider an example of two agents' preferences over four resources in Table 4.3. Let us assume that the initial allocation $A_0$ reached after Stage 1 is $A_0(1) = \{R_4, R_1\}, A_0(2) = \{R_2, R_3\}$. The corresponding negotiation tree is shown in Fig. 4.19.[5] Each node $n$ of the negotiation tree is

[5]Note that the gray nodes 3 and 15 do not belong to the negotiation tree itself and for explanation only.

**Table 4.3** An example of the agents' preferences over resources

| Allocation $A$ | $u_1(A(1))$ | $u_2(A(2))$ |
|---|---|---|
| $(\{R_1, R_2, R_3, R_4\}, \emptyset)$ | 103 | 0 |
| $(\{R_2, R_3, R_4\}, \{R_1\})$ | 91 | 63 |
| $(\{R_1, R_3, R_4\}, \{R_2\})$ | 95 | 82 |
| $(\{R_1, R_2, R_4\}, \{R_3\})$ | 92 | 64 |
| $(\{R_1, R_2, R_3\}, \{R_4\})$ | 74 | 8 |
| $(\{R_3, R_4\}, \{R_1, R_2\})$ | 88 | 90 |
| $(\{R_2, R_4\}, \{R_1, R_3\})$ | 91 | 99 |
| $(\{R_2, R_3\}, \{R_1, R_4\})$ | 65 | 67 |
| $(\{R_1, R_4\}, \{R_2, R_3\})$ | 91 | 89 |
| $(\{R_1, R_2\}, \{R_3, R_4\})$ | 59 | 64 |
| $(\{R_1, R_3\}, \{R_2, R_4\})$ | 32 | 94 |
| $(\{R_4\}, \{R_1, R_2, R_3\})$ | 83 | 104 |
| $(\{R_3\}, \{R_1, R_2, R_4\})$ | 30 | 98 |
| $(\{R_2\}, \{R_1, R_3, R_4\})$ | 58 | 103 |
| $(\{R_1\}, \{R_2, R_3, R_4\})$ | 25 | 99 |
| $(\emptyset, \{R_1, R_2, R_3, R_4\})$ | 0 | 112 |

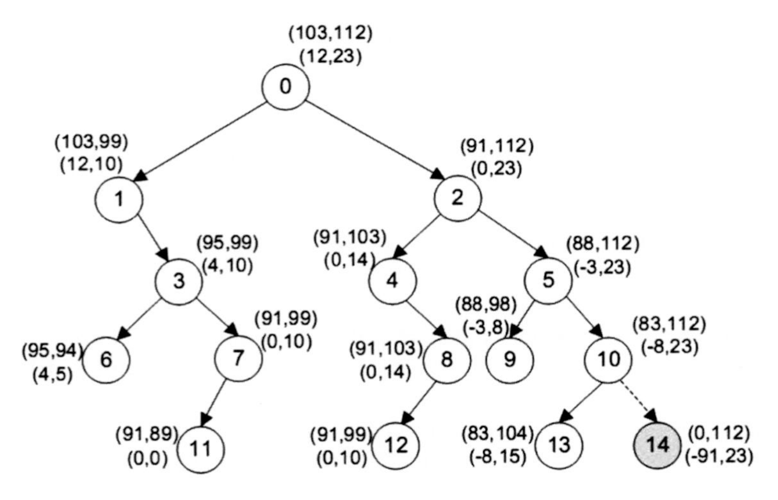


**Fig. 4.19** The negotiation tree of the example in Table 4.3

labeled with two pairs of values. The first pair of values represents the agents' best possible utilities $(BPU_1(n), BPU_2(n))$ over this node, and the second pair of values corresponds to this node's payment information $\langle P_n(1), P_n(2)\rangle$.[6] For example, considering node 5, we can have $P_5(1) = BPU_1(5) - u_1(A_0(1)) = 91 - 91 = 0$

[6]Note that each agent only knows its own best possible utility over each node in the negotiation tree. Here we show both agents' information in the same negotiation tree for illustration purpose only.

and $P_5(2) = BPU_2(5) - u_2(A_0(2)) = 103 - 89 = 14$. Each leaf node at level 4 corresponds to a complete allocation $A$ over resources.

#### 4.3.2.2 The Protocol: APSOPA

We present the three-stage protocol APSOPA in this section.

Stage 1: Initial Resource Allocation

In the first stage, the agents obtain an initial allocation $A_0$ according to the strict alternation protocol. Each agent chooses one resource from the remaining set of resources to maximize its own current utility in a myopically rational way each time. The specific allocation procedure consists of the following steps.

**Step 1** One agent $i$ is randomly chosen to pick one resource $R_i^0$ from the resource set $\mathcal{R}$, which will cause the maximum increase in its utility $u_i$. The resource set $\mathcal{R}$ is updated to $\mathcal{R} \setminus \{R_i^0\}$.
**Step 2** The other agent $j$ chooses one resource $R_j^0$ from the resource set $\mathcal{R}$, which will also bring maximum increase in its utility $u_j$. The resource set $\mathcal{R}$ is updated to $\mathcal{R} \setminus \{R_j^0\}$.
**Step 3** If $\mathcal{R} \neq \emptyset$, return to Step 1; otherwise go to Stage 2.

After Stage 1, an initial allocation $A_0$ over resources is obtained. For example, assuming that agent 1 is chosen to pick a resource first, it is easy to check that $A_0(1) = \{R_4, R_1\}, A_0(2) = \{R_2, R_3\}$ in the previous example.

Stage 2: Distributed Negotiation over Individual Resources

In this stage, the agents negotiate with each other over the allocation of individual resources in a distributed manner. The distributed negotiation between agents consists of at most $N$ steps. Each step $l$ involves the negotiation over a single resource $l$. This negotiation process can be modeled as the process of distributed generation of a negotiation tree as follows:

**Step 1** Initially the negotiation tree each agent maintains only contains the root node. The agents exchange their payment information for the root node. We define that the level of the root node is 0 and set step counter $l = 1$.
**Step 2** During step $l(l \geq 1)$, for each of the nodes in level $l - 1$, agent 1 is responsible for creating its right child node, and agent 2 is responsible for creating its left child node.
**Step 3** At the end of step $l$, the agents communicate with each other and exchange two pieces of information: (1) the child nodes it has created in current step; (2) the payment information $P_n(i)$, $i \in \{1, 2\}$, for each child node $n$ it has created in

current step. After that, each agent updates its own negotiation tree accordingly. If no child node is created at step $l$ ($l < N$), or $l = N$, then the negotiation process stops, and Stage 3 starts; otherwise step counter $l$ is increased by 1 and go to Step 2.

We know that each leaf node $n$ at level $N$ corresponds to a complete allocation $A_n$. A rational agent $i$ is willing to create a leaf node $n$ if and only if accepting the deal $(A_0, A_n)$ is an individually rational decision for itself. That is,

$$u_i(A_n(i)) + p(i) \geq u_i(A_0(i)) \tag{4.6}$$

where $p(i)$ is the payment of agent $i$. It is obvious that $p(i) = P_n(j)$ $(i \neq j)$, since the payment of agent $i$ comes from or goes to agent $j$ only. If $p(i)$ is positive, it represents the maximum payment that agent $i$ can receive under allocation $A_n$; otherwise, it is the minimum payment that agent $i$ has to pay in allocation $A_n$.

When it comes to determine whether to create the child nodes at level $l < N$, the agents cannot make their decisions based on the above decision rule since these child nodes correspond to partial allocations only. Intuitively an agent is willing to create a child node at level $l < N$ if and only if its maximum utility in the created child node is not less than its original utility under allocation $A_0$. Otherwise, any individually rational agent would have no incentive to do that. Formally agent $i$ has the incentive to create its responsible child node $n_c$ of node $n$ at level $l < N$ if and only if the following condition is satisfied.

$$BPU_i(n_c) + p(i) \geq u_i(A_0(i)) \tag{4.7}$$

where, based on the definition of payment information, we have $p(i) = P_{n_c}(j) = P_n(j)$, $(i \neq j)$.

During the negotiation in this stage, for each non-leaf node $n$ created by agent $i$, only $P_n(i)$ needs to be communicated to agent $j$. The value of $P_n(j)$ is already known to agent $i$ in the previous step which equals to $P_{n'}(j)$, where node $n'$ is the parent node of node $n$. Therefore, the agents exchange information with each other only once at the end of each step. Besides, throughout the negotiation, for each node, agent 1 only determines whether to create its right child node, and agent 2 only determines whether to create its left child node. This is because agent 1 always has the incentive to create the left child node $n_l$ of node $n$ as long as agent 2 does and vice versa.

This can be illustrated using induction as follows. Since node $n$ has been created in previous step, it means that it is an individually rational decision for agent 2, and thus we have $BPU_2(n) + p(2) \geq u_2(A_0(2))$ by letting $p(2) = -P_n(2)$. When it comes to determine whether it is individually rational for agent 2 to create node $n_r$, we only need to check whether $BPU_2(n_r) + p(2) \geq u_2(A_0(2))$ is satisfied or not. Since we know that $BPU_2(n_r) = BPU_2(n)$ and also $p(2) = -p(1) = -P_n(2)$, we can see that the previous condition is satisfied, and surely agent 2 has the incentive to create node $n_r$. The case for agent 2 is similar and is omitted here.

For example, consider the rightmost path ($0 \rightarrow 2 \rightarrow 6 \rightarrow 11 \rightarrow 15$) of the negotiation tree in Fig. 4.19, where all the nodes of this path are right child nodes created by agent 1. We can see that agent 2's best possible utility and payment information of the nodes along this path are unchanged. For each existing node on this path, it is always individually rational for agent 2 to create its right child node. Thus agent 1 can make decisions by itself without asking for agent 2's permission. Finally the gray leaf node 15 is not created by agent 1 since it violates the principle of individual rationality (see Eq. 4.6). To better illustrate when a child node will not be created, let us consider another gray node 3. We can see that agent 2 requires additional payoff of 25 to agree with this partial allocation, while the maximum payoff provided by agent 1 is only 12. Thus agent 2 has no incentive to create this node, and accordingly all potential allocations corresponding to those leaf nodes within the subtree of node 3 are eliminated.

At the end of this stage, the agents get a set $\mathcal{C}$ of candidate allocations over which both agents have the incentive to make the deal, which corresponds to the set of leaf nodes at level $N$ of the negotiation tree.

### Stage 3: Reaching Socially Optimal Allocation

In this stage, the agents negotiate with each other over the set $\mathcal{C}$ of candidate allocations in a way similar to Rubinstein's alternating-offer model [39], which involves the following steps:

**Step 1** One agent is randomly chosen to take the role of a negotiator. The initial set of candidate allocations is $\mathcal{C}$.

**Step 2** The negotiator (e.g., agent $i$) proposes a deal $(A_0, A)$ from the set $\mathcal{C}$, which brings it the highest utility among all candidate deals, and also $u_i(A(i)) + p(i) > u_i(A_0(i))$. If such a deal cannot be found, then go to Step 3 directly. Otherwise, it sends the other agent $j$ the negotiation information, which consists of two elements: the deal $(A_0, A)$ and the payment value $p(i)$. The set $\mathcal{C}$ is updated to $\mathcal{C} \setminus \{A\}$.

**Step 3** If agent $i$ does not propose any deal in Step 2 and agent $j$ ($j \neq i$) also did not propose any deal when it was the negotiator last time, or $\mathcal{C} = \emptyset$, the protocol ends. Otherwise, agent $j$ determines whether to reject the deal $(A_0, A)$. If the deal is accepted, then both agents take the new allocation $A$ as their current allocation. Let agent $j$ be the negotiator and go to Step 2.

Let us use the example in Fig. 4.19 to illustrate how the agents negotiate in this stage. We have known that $A_0(1) = \{R_1, R_4\}$ and $A_0(2) = \{R_2, R_3\}$ after Stage 1, and also $\mathcal{C} = \{12, 13, 14\}$ after Stage 2. The corresponding allocations of the leaf nodes in $\mathcal{C}$ are $A_{12}(1) = \{R_1, R_4\}, A_{12}(2) = \{R_2, R_3\}, A_{13}(1) = \{R_2, R_4\}, A_{13}(2) = \{R_1, R_3\}$, and $A_{14}(1) = \{R_4\}, A_{14}(2) = \{R_1, R_2, R_3\}$, respectively. Assume that agent 1 is the negotiator in round 1 and it proposes the deal $(A_0, A_{13})$. Since we assume that the agents are altruistic-individually rational, it will ask for payment $p(1) = 1$ from agent 2.[7] It is easy to check that $(A_0, A_{13})$ with payment value $p(2) = -p(1) = -1$ is an individually rational deal for agent 2 and thus agent 2 will accept it. The resources are now reallocated according to allocation $A_{13}$, and $\mathcal{C}$ is updated to $\{12, 14\}$. Now it is agent 2's turn to propose a new deal in round 2. It proposes the deal $(A_{13}, A_{14})$ with payment value $p(2) = -4$. We can check that agent 1 will not accept this deal since its utility will be decreased. Accordingly, $\mathcal{C}$ is updated to $\{12\}$, and it is agent 1's turn again to propose in round 3. It can be checked that agent 1 will not propose any new deal in round 3, since $u_1(\{R_1, R_4\}) = u_1(\{R_2, R_4\})$ and it is the same for agent 2 in round 4. Therefore, Stage 3 terminates after four rounds of negotiations, and the final allocation reached between the agents is $A_f(1) = \{R_2, R_4\}$ and $A_f(2) = \{R_1, R_3\}$. We can see that this allocation is indeed socially optimal.

### 4.3.3 Convergence of APSOPA to Socially Optimal Allocation

In this section, we prove that under protocol APSOPA, the final allocation is always socially optimal as long as the agents are altruistic-individually rational. This convergence property can be proved based on the following two lemmas.

**Lemma 4.1** *The set $\mathcal{C}$ of allocations obtained at the end of Stage 2 consists of all and only the allocations satisfying the property that $\forall A_i \in \mathcal{C}$, the deal $(A_0, A_i)$ is individually rational.*

*Proof* The **only** part: Consider a leaf node (allocation) $A_i$ $(A_i \in \mathcal{C})$. Without loss of generality, assume it is created by agent 1. Since agent 1 is altruistic-individually rational, we have $u_1(A(1)) + p(1) \geq u_1(A_0(1))$, which implies that the deal $(A_0, A_i)$ is individually rational for agent 1. Besides, according to the analysis in Sect. 4.3.2.2, we know that the deal $(A_0, A_i)$ is always individually rational for agent 2 as long as it is individually rational for agent 1. Thus every deal $(A_0, A_i)$ $(A_i \in \mathcal{C})$ is individually rational.

The **all** part: Suppose that there exists one allocation $A' \notin \mathcal{C}$, but the corresponding deal $(A_0, A')$ is individually rational. The allocation $A'$ must correspond to a leaf node $n'$ in the negotiation tree which is not created by the agents in Stage 2. Since the deal $(A_0, A')$ is individually rational, we have $u_1(A') + p(1) \geq u_1(A_0)$ and $u_2(A') + p(2) \geq u_2(A_0)$. Considering its parent node $n'_f$ in previous level, there exist two possible conditions as follows:

- If node $n'$ is the left child node of $n'_f$, for node $n'_f$, we can have $u_1(A') + p(1) \geq u_1(A_0)$ and $u_2(A' + \{R_N\}) + p(2) \geq u_2(A') + p(2) \geq u_2(A_0)$. It means that the agents should have the incentives to create this node $n'_f$.

[7] Recall that we assume that the agent's utilities are integers only and the utility's minimum unit is 1. Since the agents are altruistic-individually rational, and also $u_1(A_0(1)) = u_i(A_{13}(1))$, agent 1 will ask for a payment of $p(1) = 1$ to have the incentive to propose the allocation $A_{13}$.

- If node $n'$ is the right child node of $n'_f$, then we can have $u_1(A' + \{R_N\}) + p(1) \geq u_1(A') + p(1) \geq u_1(A_0)$ and $u_2(A') + p(2) \geq u_2(A_0)$. It also indicates that node $n'_f$ should have been created in Stage 2.

We can repeat reasoning in this way until the root node to show that all nodes on the path should have been created during the negotiation tree generation process. Therefore, the leaf node $n'$ also should have been created by the agents at the end of Stage 2 provided that the agents are altruistic-individually rational. Therefore, we reach a contradiction here, and we can have the conclusion that there does not exist a leaf node which is not created after Stage 2, but its corresponding allocation is individually rational for both agents with respect to allocation $A_0$. □

**Lemma 4.2** *Given the initial allocation $A_0$ and the set of candidate allocations $\mathcal{C} = \{A \mid (A_0, A) \text{is individually rational}\}$, the allocation $A_f$ the agents finally reach through Stage 3 is guaranteed to be socially optimal, i.e., $\sum_{i \in \mathcal{A}} u_i(A_f(i))$ is maximized.*

*Proof* To prove this lemma, first we need the following property of an individually rational deal: a deal $(A_1, A_2)$ is individually rational iff $sw(A_1) \leq sw(A_2)$. This property was first proved in [32], and thus we omit the proof here.

Since $\mathcal{C} = \{A \mid (A_0, A) \text{ is individually rational}\}$, for each allocation $A \in \mathcal{C}$, we can have $sw(A) \geq sw(A_0)$ based on the above property. Let $A_f = \arg\max_{A \in \mathcal{C}} sw(A)$, it is easy to see that this allocation is socially optimal. To prove that the final allocation is socially optimal, we only need to show that allocation $A_f$ will always be reached and the agents will never accept any other allocation thereafter.

Suppose that the allocation in round $t$ of Stage 3 is $A_t$ and $sw(A_t) < sw(A_f)$. At least one agent can have the incentive to propose the deal $(A_t, A_f)$ with the increase of one unit of its utility[8]; otherwise we get a contradiction here since the condition $sw(A_t) < sw(A_f)$ is violated. Besides, since $sw(A_f) > sw(A_t)$, and also the payment function satisfies that the agent proposing the deal only asks for the increase of one unit of its utility, the deal $(A_t, A_f)$ must be individually rational (Definition 4.3), and it will be accepted by both agents. If other deals are proposed instead of $(A_t, A_f)$ in round $t$, we can know that the deal $(A_t, A_f)$ will eventually be proposed and accepted by the agents by applying the same induction repeatedly.

After allocation $A_f$ is reached, no agent will have the incentive to propose any other allocation. This can be proved by contradiction. Without loss of generality, suppose that agent 1 has the incentive to propose another deal $(A_f, A')$, there must exist a payment function $p$ such that $u_1(A'(1)) + p(1) > u_1(A_f(1))$ and $u_2(A'(2)) + p(2) \geq u_2(A_f(2))$. Since $p(1) + p(2) = 0$, we get $u_1(A') + u_2(A') > u_1(A_f) + u_2(A_f)$, that is, $sw(A') > sw(A_f)$, where we get a contradiction with the fact that $A_f =$

[8]Note that this is based on the assumption that the agents are altruistic-individually rational. This assumption is important to prevent that the socially optimal allocation may be discarded during negotiation. For example, consider a deal $(A_t, A_{t+1})$ in which $A_{t+1}$ is the socially rational allocation, and $u_1(A_t(1)) = 10$, $u_2(A_t(2)) = 6$, $u_1(A_{t+1}(1)) = 15$, and $u_2(A_{t+1}(2)) = 2$. Without this assumption, agent 1 may propose the deal $(A_t, A_{t+1})$ with $p(1) = 3$, and accordingly, agent 2 will reject this offer since its utility is decreased.

$\arg\max_{A \in C} sw(A)$. Hence after allocation $A_f$ is reached, the negotiation process in Stage 3 will continue for at most two rounds. Overall we get the conclusion that the final allocation reached is always the socially optimal one $A_f$. □

Based on Lemmas 4.1 and 4.2, we can easily get the conclusion that the final allocation reached by altruistic-individually rational agents using APSOPA is socially optimal.

**Theorem 4.2** *The final allocation reached by the altruistic-individually rational agents under the protocol APSOPA is socially optimal.*

### *4.3.4 Experimental Evaluation*

In this section, we experimentally evaluate the performance of the APSOPA protocol. We consider different problem sizes by varying the number of resources $N$. For each $N$, we generate 200 random examples of both agents' preferences over resources. We execute the negotiation process under the APSOPA protocol, and the average performance of the APSOPA protocol for different problem sizes is obtained for evaluation.

Firstly, we observe that the simulation results show that the final allocations reached by the agents using APSOPA are always socially optimal, which is in accordance with the theoretical results in Sect. 4.3.3. Secondly, we experimentally evaluate the complexity of the APSOPA protocol and show its efficiency from the following two perspectives: communication complexity and computational complexity. Communication complexity of a negotiation process is termed in the work of Endriss and Maudet [28], which focuses on investigating the following question:

- How many deals are required among the agents before finally reaching the socially optimal allocation?

The negotiation over a particular deal typically involves two-way communication between the agents, i.e., one agent proposes the deal and the other agent makes a reply saying yes or no. If we take an individual deal as the primitive and abstract away the complexity behind it, it is reasonable to use the number of deals required during negotiation to evaluate the communication complexity of the negotiation process. Under the APSOPA protocol, the agents need to negotiate over the candidate allocations obtained from Stage 2, i.e., the leaf nodes of the negotiation tree at level $N$. Thus the number of deals required to reach socially optimal allocation in Stage 3 is at most the number of leaf nodes of the negotiation tree at level $N$. Besides, in APSOPA, the agents also need to communicate with each other for at most $N$ times in Stage 2. Here we simply count each two-way communication in Stage 2 as one deal. For example, consider the previous example in Fig. 4.19. It contains three leaf nodes, and the height of the negotiation tree is 4; thus, the maximum total number of deals required is $4 + 3 = 7$.

We evaluate the communication complexity of APSOPA comparing with the following two baselines:

**Case 1** Stage 1 is removed, i.e., agents start the negotiation following the protocol in Stages 2 and 3 with a random initial allocation.

**Case 2** Stage 2 is removed, i.e., the agents start the negotiation process in Stage 3 over the original allocation space based on the initial allocation $A_0$ obtained in Stage 1.

The comparison results are shown in Table 4.4, in which only approximately 30 % and 12 % of the deals are required using the complete APSOPA protocol compared with the above two cases, respectively. We make the following observations. First, we can see that the choice of the initial allocation can have significant effects on the communication cost, and it is worthwhile to carefully pick a good initial allocation before entering Stage 2 in the cost of the slight increasing of the computational cost. Second, the simulation results also show the efficiency of Stage 2 in reducing the size of the set of candidate deals to negotiate over in Stage 3, since approximately 88 % of irrelevant deals are eliminated during Stage 2. Overall we can see that both stages are efficient and important in reducing the communication cost of the negotiation process.

Next we consider the performance of the APSOPA protocol in terms of its computational cost. We evaluate the computational cost by considering the number of allocations that each agent has to search before the socially optimal one is reached [29]. In APSOPA protocol, the computational complexity of an agent is mainly involved in Stage 2. An agent needs to consider a new allocation and calculate its preference over this allocation whenever a new node is created in Stage 2. We experimentally compare the average number of allocations that each agent searches using APSOPA with that under protocols like one-step monotonic concession protocol [34] under complete information. Table 4.5 shows the experimental comparison results for different number of resources. From Table 4.5, we can see that the search effort of each agent is greatly reduced using the APSOPA protocol, and thus the computational complexities of the agents become much lower.

**Table 4.4** Average communication complexity of APSOPA

| No. of resources | Average no. of deals using APSOPA | % of deals for case 1 | % of deals for case 2 |
|---|---|---|---|
| 4 | 6 | 60 | 40 |
| 8 | 33 | 32.3 | 12.9 |
| 12 | 359 | 24.6 | 12.4 |
| 14 | 1404 | 22.5 | 10.2 |
| 16 | 5120 | 21.4 | 12.2 |

**Table 4.5** Average computational complexity of APSOPA

| No. of resources | Average no. of allocations | % of allocations searched by each agent using APSOPA |
|---|---|---|
| 4 | 10 | 62.5 |
| 8 | 157 | 61.3 |
| 12 | 2587 | 63.2 |
| 14 | 10,486 | 64.0 |
| 16 | 41,578 | 59.6 |

## References


1. Hoen PJ, Tuyls Kl, Panait L, Luke S, Poutre JAL (2005) An overview of cooperative and competitive multiagent learning. In: Proceedings of first international workshop on learning and adaption in multi-agent systems, Utrecht, pp 1–46
2. Hao JY, Leung HF (2013) Reinforcement social learning of coordination in cooperative multi-agent systems(extended abstract). In: Proceedings of AAMAS'13, Saint Paul, pp 1321–1322
3. Hao JY, Leung HF (2013) The dynamics of reinforcement social learning in cooperative multiagent systems. In: Proceedings of IJCAI 13, Beijing, pp 184–190
4. Matlock M, Sen S (2007) Effective tag mechanisms for evolving coordination. In: Proceedings of AAMAS'07, Toronto, p 251
5. Hao JY, Leung HF (2011) Learning to achieve social rationality using tag mechanism in repeated interactions. In: Proceedings of ICTAI'11, Washington, DC, pp 148–155
6. Panait L, Luke S (2005) Cooperative multi-agent learning: the state of the art. Auton Agents Multi-Agent Syst 11(3):387–434
7. Fulda N, Ventura D (2007) Predicting and preventing coordination problems in cooperative learning systems. In: Proceedings of IJCAI'07, Hyderabad
8. Matignon L, Laurent GJ, Le For-Piat N (2012) Independent reinforcement learners in cooperative Markov games: a survey regarding coordination problems. Knowl Eng Rev 27:1–31
9. Claus C, Boutilier C (1998) The dynamics of reinforcement learning in cooperative multiagent systems. In: Proceedings of AAAI'98, Madison, pp 746–752
10. Lauer M, Rienmiller M (2000) An algorithm for distributed reinforcement learning in cooperative multi-agent systems. In: Proceedings of ICML'00, Stanford, pp 535–542
11. Kapetanakis S, Kudenko D (2002) Reinforcement learning of coordination in cooperative multiagent systems. In: Proceedings of AAAI'02, Edmonton, pp 326–331
12. Matignon L, Laurent GJ, Le For-Piat N (2008) A study of FMQ heuristic in cooperative multi-agent games. In: AAMAS'08 workshop: MSDM, Estoril, pp 77–91
13. Panait L, Sullivan K, Luke S (2006) Lenient learners in cooperative multiagent systems. In: Proceedings of AAMAS'06, Utrecht, pp 801–803
14. Wang X, Sandholm T (2002) Reinforcement learning to play an optimal nash equilibrium in team Markov games. In: Proceedings of NIPS'02, Vancouver, pp 1571–1578
15. Brafman RI, Tennenholtz M (2004) Efficient learning equilibrium. Artif Intell 159:27–47
16. Watkins CJCH, Dayan PD (1992) Q-learning. Mach Learn 8:279–292
17. Melo FS, Veloso M (2009) Learning of coordination: exploiting sparse interactions in multiagent systems. In: Proceedings of AAMAS'09, Budapest, pp 7730–780
18. Sen S, Airiau S (2007) Emergence of norms through social learning. In: Proceedings of IJCAI'07, Hyderabad, pp 1507–1512
19. Fudenberg D, Levine DK (1998) The theory of learning in games. MIT, Cambridge

20. Hales D, Edmonds B (2003) Evolving social rationality for mas using "tags". In: Proceedings of AAMAS'03. ACM, New York, pp 497–503
21. Sen S, Arora N, Roychowdhury S (1998) Using limited information to enhance group stability. Int J Hum-Comput Stud 48:69–82
22. Matlock M, Sen S (2009) Effective tag mechanisms for evolving coperation. In: Proceedings of AAMAS'09, Budapest, pp 489–496
23. Hogg LM, Jennings NR (1997) Socially rational agents. In: Proceedings of AAAI fall symposium on socially intelligent agents, Providence, pp 61–63
24. Hogg LMJ, Jennings NR (2001) Socially intelligent reasoning for autonomous agents. IEEE Trans SMC Part A Syst Hum 31:381–393
25. Chao I, Ardaiz O, Sanguesa R (2008) Tag mechanisms evaluated for coordination in open multi-agent systems. In: Proceedings of 8th international workshop on engineering societies in the agents world, Athens, pp 254–269
26. Chevaleyre Y, Dunne PE et al (2006) Issues in multiagent resource allocation. Informatica 30:3–31
27. Chevaleyre Y, Endriss U, Maudet N (2006) Tractable negotiation in tree-structured domains. In: Proceedings of AAMAS'06, Hakodate, pp 362–369
28. Endriss U, Maudet N (2005) On the communication complexity of multilateral trading: extended report. Auton Agents Multi-Agent Syst 11:91–107
29. Saha S, Sen S (2007) An efficient protocol for negotiation over multiple indivisible resources. In: Proceedings of IJCAI'07, Hyderabad, pp 1494–1499
30. Maly K, Overstreet C, Qiu X, Tang D (1988) Dynamic bandwidth allocation in a network. In: Proceedings of the ACM symposium on communications architectures and protocols, Stanford
31. Gomoluch J, Schroeder M (2003) Market-based resource allocation for grid computing: a model and simulation. In: Proceedings of MGC'03, Rio de Janeiro, pp 211–218
32. Endriss U, Maudet N, Sadri F, Toni F (2003) On optimal outcomes of negotiation over resources. In: Proceedings of AAMAS'03, Melbourne
33. Chevaleyre Y, Endriss U, Maudet N (2010) Simple negotiation schemes for agents with simple preferences: sufficiency, necessity and maximality. Auton Agents Multi-Agent Syst 20(2):234–259
34. Rosenschein JS, Zlotkin G (1994) Rules of encounter. MIT, Cambridge
35. Brams SJ, Taylor AD (2000) The win-win solution: guaranteeing fair shares to everybody. W.W.Norton and Company, New York
36. Endriss U, Maudet N (2003) Welfare engineering in multiagent systems. In: Engineering societies in the agents world IV. Springer, Berlin, pp 93–106
37. Arrow KJ, Sen AK, Suzumura K (2002) Handbook of social choice and welfare. North-Holland, Amsterdam
38. Brams SJ, Taylor AD (1996) Fair division: from cake-cutting to dispute resolution. Cambridge University Press, Cambridge
39. Rubinstein A (1982) Perfect equilibrium in a bargaining model. Econometrica 50(1):97–110

# Chapter 5
# Individual Rationality in Competitive Multiagent Systems

In competitive MASs, each individual agent is usually interested in maximizing its personal benefits only, which may have conflicts with the utility of others and the overall system as well. Thus, one natural research direction in competitive MASs is to consider how an agent can learn to obtain as much utility as possible against different opponents based on its local information. Another important question is raised from the system designer's perspective, i.e., how can the selfish agents be incentivized to coordinate their behaviors to maximize the system-level performance (i.e., maximizing social optimality)? In this chapter, we focus on the first research direction by considering an important competitive multiagent interaction scenario: bilateral negotiation [1]. The second research direction will be the focus of Chap. 6.

## 5.1 Introduction

Negotiation is a common and important approach to resolve conflicts and reach agreements between different parties in our daily life. The topic of negotiation has been widely studied in various areas, such as decision and game theory, economics, and social science [2]. Automated negotiation techniques can, to a large extent, alleviate the efforts of human and also facilitate human in reaching better negotiation outcomes by compensating for the limited computational abilities of humans when they are faced with complex negotiations. Until now, a lot of automated negotiation strategies and mechanisms have been proposed in different negotiation scenarios [3–6].

The major difficulty in designing automated negotiation agent is how to achieve optimal negotiation results given incomplete information on the negotiating partner. The negotiation partner usually keeps its negotiation strategy and its preference as its private information to avoid exploitations. A lot of research efforts have been devoted to better understand the negotiation partner by either estimating



J. Hao, H.-f. Leung, *Interactions in Multiagent Systems: Fairness, Social Optimality and Individual Rationality*, DOI 10.1007/978-3-662-49470-7_5

the negotiation partner's preference profile [5, 7, 8] or predicting its decision function [6, 9]. On one hand, with the aid of different preference profile modeling techniques, the negotiating agents can get a better understanding of their negotiating partners and thus increase their chances of reaching mutually beneficial negotiation outcomes. On the other hand, effective strategy prediction techniques enable the negotiating agents to maximally exploit their negotiating partners and thus receive as much benefit as possible from negotiation. However, in most of previous work, the negotiating agents are usually assumed to be situated in a negotiation environment in that strong limitations are put on the negotiation scenario or the negotiation opponent. For example, some work assumes that the agents negotiate over a single item only; however practical negotiation scenarios usually involve multiple items to negotiate over. There also exists some work that assumes that the negotiation agents can have access to their opponents' (partial) preferences. This can be unrealistic especially in multi-issue negotiation scenarios in which the preferences of different agents may vary significantly, and the agents usually would not like to disclose their preferences to avoid being exploited. Another assumption commonly adopted is that the negotiation opponent is limited to choose from a specific set of simple strategies, e.g., time-dependent or behavior-dependent tactics. Those strategies designed under this assumption may not work well against other negotiation partners with more complex state-of-the-art strategies.

To this end, in recent years a number of advanced negotiation strategies taking advantage of existing techniques have been proposed, and agents employing these strategies have participated in *automated negotiating agents competition (ANAC)* [10, 11]. The ANAC provides a negotiation platform which enables different negotiation agents to be evaluated against a wide range of opponents within a realistic negotiation environment. During the past 3 years, dozens of state-of-the-art negotiation strategies have been extensively evaluated in a variety of multi-issue negotiation scenarios, and valuable insights have been obtained in terms of the advantages and disadvantages of different techniques, e.g., the efficacy of different acceptance conditions [12]. It is still an open and interesting problem to design more efficient automated negotiation strategies against a variety of negotiating opponents in different negotiation domains.

In this chapter, we describe an adaptive negotiation *ABiNeS* strategy for automated agents to negotiate in bilateral multi-issue negotiation environments following the settings adopted in ANAC 2012 [13].[1] Bilateral multi-issue negotiations surround people's daily life and have received lots of attention in the negotiation literature. During negotiation, both the agents' negotiation strategies and preference profiles are their private information, and for each agent the only available information about the negotiating partner is its past negotiation moves. Considering the diversity of the available negotiation strategies that the negotiating agents can adopt, it is usually very difficult (or impossible) to predict which specific strategy the negotiating partner is using based on this limited information. To effectively

[1]This negotiation strategy won the champion of ANAC 2012.

cope with different types of opponents, the *ABiNeS* negotiation agent introduces the concept of nonexploitation point $\lambda$ to adaptively adjust the degree that an *ABiNeS* agent exploits its negotiating opponent. The value of $\lambda$ is determined by the characteristics of the negotiation scenario and the concessive degree of the negotiating partner, which is estimated based on the negotiation history. Besides, to maximize the possibility that the offer the *ABiNeS* agent proposes will be accepted by its negotiating partner, it can be useful to make predictions on the preference profile of the negotiating partner. Instead of explicitly modeling the negotiation partner's preference profile, a reinforcement learning-based approach is adopted in *ABiNeS* agent to determine the optimal proposal for the negotiating partner based on the current negotiation history.

We evaluate the performance of the *ABiNeS* agent compared with a number of state-of-the-art negotiation strategies from two different perspectives: *efficiency* in terms of the average payoff obtained under a particular negotiation tournament and *robustness* in terms of how likely the agents have the incentive to adopt our strategy rather than other strategies. The *efficiency* evaluation is conducted under the negotiation tournament setting following ANAC 2012 using GENIUS[2] [14] platform. Simulation results show that the *ABiNes* agent can make more effective exploitations against a variety of negotiation partners and thus obtain higher average payoffs during negotiation tournaments. We propose adopting existing model checking techniques [15] to perform the *robustness* analysis, which, as will be shown later, is proved to be a very convenient and efficient technique to perform similar types of robustness analysis tasks. According to the *robustness* analysis, the *ABiNes* strategy is shown to be very robust against other state-of-the-art strategies under different negotiation contexts.

The remainder of Chap. 5 is organized as follows: First we give a description of negotiation model in Sect. 5.2. In Sect. 5.3, the negotiation *ABiNes* strategy we propose is introduced. In Sect. 5.4, we evaluate the negotiation *efficiency* and *robustness* of *ABiNes* with the state-of-the-art agents under different negotiation contexts.

## 5.2 Negotiation Model

In this section, we describe the negotiation model which follows the settings adopted in ANAC 2012 [13]. We focus on bilateral negotiations, i.e., negotiations between two agents. Specifically, the alternating-offer protocol is adopted to regulate the interactions between the negotiating agents, in which the agents take turns to exchange proposals. For each negotiation scenario, both agents can negotiate over multiple issues (items), and each item can have a number of different values. Let

---

[2]GENIUS is short for General Environment for Negotiation with Intelligent multipurpose Usage Simulation.

us denote the set of items as $\mathcal{M}$ and the set of values for each item $m_i \in \mathcal{M}$ as $\mathcal{V}_i$.[3] We define a negotiation outcome $\omega$ as a mapping from every item $m_i \in \mathcal{M}$ to a value $v \in \mathcal{V}_i$, and the negotiation domain is defined as the set $\Omega$ of all possible negotiation outcomes. For each negotiation outcome $\omega$, we use $\omega(m_i)$ to denote the corresponding value of the item $m_i$ in the negotiation outcome $\omega$. We assume that the knowledge of the negotiation domain is known to both agents beforehand and is not changed during the whole negotiation session.

For each negotiation outcome $\omega$, different agents may have different preferences. Here we assume that each agent $i$'s preference can be modeled by a utility function $u_i$ such that $\forall \omega \in \Omega$, it is mapped into a real-valued number in the range of [0,1], i.e., $u_i(\omega) \in [0, 1]$. In practical negotiation environments, there usually associates a certain cost with each negotiation. To take this factor into consideration, a real-time deadline is imposed on the negotiation process and each agent's actual utilities over the negotiation outcomes are decreased by a discounting factor $\delta$ over time. Following the setting adopted in ANAC 2012, each negotiation session is allocated 3 min, which is normalized into the range of [0,1], i.e., $0 \leq t \leq 1$. Formally, if an agreement is reached at time $t$ before the deadline, each agent $i$'s actual utility function $U_i^t(\omega)$ over this mutually agreed negotiation outcome $\omega$ is defined as follows:

$$U_i^t(\omega) = u_i(\omega)\delta^t. \tag{5.1}$$

If no agreement is reached by the deadline, each agent $i$ will obtain a utility of $ru_i^0\delta$, where $ru_i^0$ is agent $i$'s private reservation value in the negotiation scenario. The agents will also obtain their corresponding reservation values if the negotiation is terminated before the deadline. Note that the agents' actual utilities over their reservation values are also discounted by the discounting factor $\delta$ over time $t$. We assume that the agents' preference information and their reservation values are private and cannot be accessed by their negotiating partners.

The interaction between the negotiation agents is regulated by the alternating-offer protocol, in which the agents are allowed to take turns to exchange proposals. During each encounter, if it is agent $i$'s turn to make a proposal, it is allowed to make a choice from the following three options:

**Accept the offer from its negotiating partner.**
In this case, the negotiation ends and an agreement is reached. Both agents will obtain the corresponding utilities according to Eq. 5.1, where $\omega$ is the negotiation outcome that they mutually agree with.

**Reject and propose a counteroffer to its negotiating partner.**
In this case, the negotiation process continues and it is its negotiating partner's turn to make a counterproposal next time provided that the deadline is not reached yet.

[3] Here $\mathcal{V}_i$ can be either discrete values or continuous real values.

**Terminate the negotiation.**
In this case, the negotiation terminates and each agent $i$ gets its corresponding utility based on its private reservation value with the initial value of $ru_i^0$. Note that their actual utilities in this case are also decreased over time by the same discounting factor $\delta$, i.e., $U_i^t = ru_i^0 * \delta^t$.

Overall, the negotiation process terminates when either of the following conditions is satisfied: (1) the deadline is reached (*End*); (2) an agent chooses to terminate the negotiation before reaching the deadline (*Terminate*); (3) an agent chooses to accept the negotiation outcome proposed by its negotiating partner (*Accept*).

For each negotiation session between two agents $A$ and $B$, let $x_{A \to B}^t$ denote the negotiation outcome proposed by agent $A$ to agent $B$ at time $t$. Naturally a negotiation history $H_{A \leftrightarrow B}^t$ between agent $A$ and $B$ until time $t$ can be represented as follows:

$$H_{A \leftrightarrow B}^t := (x_{p_1 \to p_2}^{t_1}, x_{p_3 \to p_4}^{t_2}, \ldots, x_{p_n \to p_{n+1}}^{t_n}) \tag{5.2}$$

where

- $t_k \leq t_l$ for $k \leq l$, i.e., the negotiation outcomes are ordered over time, and also $t_n \leq t$.
- $p_k = p_{k+2} \in \{A, B\}$, i.e., the negotiation process strictly follows the alternating-offer protocol.

Similarly, we denote the negotiation history during a certain period of time (between time $t_1$ and $t_2$) as $H_{A \leftrightarrow B}^{t_1 \to t_2}$. As previously mentioned, we know that a negotiation session between two agents $A$ and $B$ will terminate either when one agent chooses an action from the set $\{Accept, Terminate\}$ or the deadline is reached. Thus the last element of a complete negotiation history can be either one of the elements from the set $S = \{Accept, Terminate, End\}$. A negotiation history by time $t$ is active if its last element is not equal to any element in $S$.

## 5.3 *ABiNeS*: An Adaptive Bilateral Negotiating Strategy

In this section, we describe the adaptive negotiating *ABiNeS* strategy in details. For the ease of description, we refer to the *ABiNeS* agent as agent $A$ and its negotiating partner as agent $B$ in the following descriptions. The *ABiNeS* strategy mainly consists of four basic decision components. The first component is *acceptance-threshold (AT)* component and it is responsible for determining the *ABiNeS* agent's minimum acceptance threshold $l_A^t$ at time $t$. The second component is *next-did (NB)* component whose function is to determine the negotiation outcome $x_{A \to B}^t$ that the *ABiNeS* agent proposes at time $t$. The third component, *acceptance-condition (AC)* component, is used for determining whether to accept the current proposal from agent $B$ or not. Given a negotiation history $H_{A \leftrightarrow B}^t$, its acceptance threshold

**Algorithm 4** Overall flow of the *ABiNeS* strategy

1: **for** each negotiation history $H^t_{A\leftrightarrow B}$: $= (x^{t_1}_{p_1\rightarrow p_2}, x^{t_2}_{p_3\rightarrow p_4}, \ldots, x^{t_m}_{p_n\rightarrow p_{n+1}})$ at current time $t$ **do**
2: Determine the acceptance threshold $l^t$ and the negotiation outcome $x^t_{A\rightarrow B}$ to be proposed to the negotiating partner $B$, and $ru^t_A = ru^0_i * \delta^t$.
3: **if** $H^t_{A\leftrightarrow B}$ is empty **then**
4: **if** $TC(H^t_{A\leftrightarrow B}, l^t_A, x^t_{A\rightarrow B}, ru^0_A)$ is false **then**
5: Propose the negotiation outcome $x^t_{A\rightarrow B}$ to the negotiating partner.
6: **else**
7: Terminate the negotiation.
8: **end if**
9: **else**
10: **if** $AC(H^t_{A\leftrightarrow B}, l^t_A, x^t_{A\rightarrow B})$ is true and $TC(H^t_{A\leftrightarrow B}, l^t_A, x^t_{A\rightarrow B}, ru^0_A)$ is false **then**
11: Accept the offer.
12: **else if** $AC(H^t_{A\leftrightarrow B}, l^t_A, x^t_{A\rightarrow B})$ is false and $TC(H^t_{A\leftrightarrow B}, l^t_A, x^t_{A\rightarrow B}, ru^0_A)$ is true **then**
13: Terminate the negotiation.
14: **else if** $AC(H^t_{A\leftrightarrow B}, l^t_A, x^t_{A\rightarrow B})$ is true and $TC(H^t_{A\leftrightarrow B}, l^t_A, x^t_{A\rightarrow B}, ru^0_A)$ is true **then**
15: **if** $U^t_A(x^{t_m}_{p_n\rightarrow p_{n+1}}) > ru^t_A$ **then**
16: Accept the offer.
17: **else**
18: Terminate the negotiation.
19: **end if**
20: **else**
21: Propose the negotiation outcome $x^t_{A\rightarrow B}$ to the negotiating partner.
22: **end if**
23: **end if**
24: **end for**

$l^t_A$, and its negotiation outcome $x^t_{A\rightarrow B}$ to propose at time $t$, the AC component returns a boolean value indicating whether to accept the offer or not, which is denoted as $AC(H^t_{A\leftrightarrow B}, l^t_A, x^t_{A\rightarrow B})$. The last component is *Termination-Condition (TC)* component. It is responsible for determining whether to terminate the current negotiation or not. Similar to the AC component, given a negotiation history $H^t_{A\leftrightarrow B}$, its acceptance threshold $l^t_A$, its negotiation outcome $x^t_{A\rightarrow B}$ to propose at time $t$, and its reservation utility $ru^0_A$, the TC component returns a boolean value indicating whether to terminate the negotiation or not, which is denoted as $TC(H^t_{A\leftrightarrow B}, l^t_A, x^t_{A\rightarrow B}, ru^0_A)$.

The overall description of the negotiation *ABiNeS* strategy based on the above basic elements is presented in Algorithm 4. At time $t$ the acceptance threshold and the next-round negotiation outcome are calculated first based on the AT and NB components, respectively. If the *ABiNeS* agent is the first one to make a proposal, then it is faced with two options: either proposing a negotiation outcome to agent $B$ or terminating the negotiation (Lines 2–8). Otherwise, the *ABiNeS* agent makes a choice among three options: proposing a negotiation outcome to agent $B$, choosing to terminate the negotiation, and accepting the offer from agent $B$ (Lines 10–22). The decisions on whether to accept the offer or terminate the negotiation are determined by the AC and TC components, respectively. We introduce each decision component in details in the following sections.

### 5.3.1 Acceptance-Threshold (AT) Component

The AT component is responsible for determining the acceptance threshold $l$ of the *ABiNeS* strategy during negotiation. The value of the acceptance threshold reflects the agent's current concession degree and should be adaptively adjusted based on the past experience and the characteristic of the negotiation environment. Besides, it also has explicit influence on the decision-making process of the other three components, which will be introduced later.

We assume that the negotiating partner is self-interested, and it will accept any proposal when the deadline is approaching ($t \sim 1$). Therefore the acceptance threshold of an *ABiNeS* agent should be always higher than the highest utility it can obtain when $t = 1$. Specifically, at any time $t$, the acceptance threshold $l_t$ of the *ABiNeS* agent should not be lower than $u^{\max}\delta^{1-t}$, where $u^{\max}$ is its maximum utility over the negotiation domain without discounting. Since the negotiating goal is to reach an agreement which maximizes the agent's own utility as much as possible, its negotiating partner should be exploited as much as possible by setting its acceptance threshold as high as possible. One the other hand, due to the discounting effect, the actual utility the agent receives can become extremely low though its original utility over the mutually agreed negotiation outcome is high, if it takes too long for the agents to reach the agreement. In the worst case, the negotiation may end up with a break-off and each agent obtains zero utility. Thus we also need to make certain compromises to the negotiating partner, i.e., lower the acceptance threshold, depending on the type of the partner we are negotiating with. Therefore, the key problem is how to balance the trade-off between exploiting and making compromise to the negotiating partner. Toward this end, we introduce the adaptive nonexploitation point $\lambda$, which represents the specific time when we should stop exploitations on the negotiating partner. This value is adaptively adjusted based on the behavior of the negotiating partner. Specifically we propose that for any time $t < \lambda$, the *ABiNeS* agent always exploits its negotiating partner (agent $B$) by setting its acceptance threshold to a value higher than $u^{\max}\delta^{1-\lambda}$ and approaching this value until time $\lambda$ according to certain pattern of behavior. After time $\lambda$, its acceptance threshold is set to be equal to $u^{\max}\delta^{1-t}$ forever, and any proposal over which its utility is higher than this value will be accepted. Formally, the acceptance threshold $l_A^t$ of an *ABiNeS* agent at time $t$ is determined as follows:

$$l_A^t = \begin{cases} u^{\max} - (u^{\max} - u^{\max}\delta^{1-\lambda})(\frac{t}{\lambda})^{\alpha} & \text{if } t < \lambda \\ u^{\max}\delta^{1-t} & \text{otherwise} \end{cases} \tag{5.3}$$

where the variable $\alpha$ controls the way the acceptance threshold approaches $u^{\max}\delta^{1-t}$ (boulware ($\alpha > 1$), conceder ($\alpha < 1$), or linear ($\alpha = 1$)). One example showing the dynamics of the acceptance threshold with time $t$ with different value of $\lambda$ is given in Fig. 5.1.

The remaining question is how to calculate the value of nonexploitation point $\lambda$. The value of $\lambda$ is determined by the characteristics of the negotiation scenario

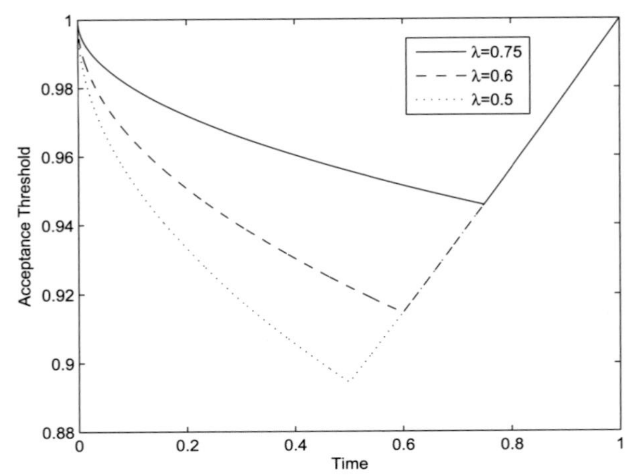


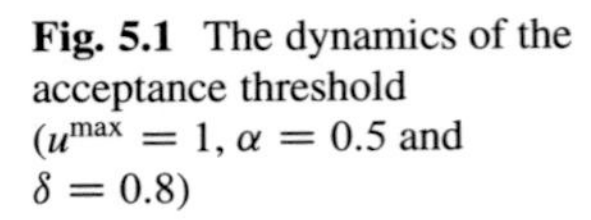
**Fig. 5.1** The dynamics of the acceptance threshold ($u^{\max} = 1, \alpha = 0.5$ and $\delta = 0.8$)

(i.e., discounting factor $\delta$) and the concession degree of the negotiating partner. The smaller the discounting factor $\delta$ is, the less actual utility we will receive as time goes by, which means more risk we are facing when we continue exploiting the negotiating partner. Therefore the value of $\lambda$ should be decreased with the decreasing of the discounting factor $\delta$. The concession degree of the negotiating partner is estimated based on its past behaviors. Intuitively, the more number of new negotiation outcomes that the negotiating partner has recently proposed, the more it is willing to make concession to end the negotiation. Specifically, the negotiation partner's concessive degree $\sigma^t$ is defined as the ratio of new negotiation outcomes it proposed within the most recent finite negotiation history $H^{t' \rightarrow t}_{A \leftrightarrow B}$. If we predict that the negotiating partner is becoming more concessive, we can take advantage of this prediction by postponing the time we stop exploitations, i.e., increasing the value of $\lambda$.

Initially the value of $\lambda$ is determined by the discounting factor $\delta$ only since we do not have any information on the negotiating partner yet. After that, it is adaptively adjusted based on the estimation of the concession degree of the negotiating partner. The overall adjustment rule of $\lambda$ during negotiation is shown in Fig. 5.2.

### 5.3.2 Next-Bid (NB) Component

The next-bid component is responsible for determining the negotiation outcome to be proposed to the negotiating partner. Given the current acceptance threshold $l^t_A$ at time $t$, any negotiation outcome over which the *ABiNeS* agent's utility is higher than $l^t_A$ can be a reasonable outcome to propose. To maximize the likelihood that the offer will be accepted by agent $B$, we need to predict the negotiation outcome $\omega_{\max}$ which can maximize its utility among the set $C$ of candidate negotiation outcomes

---

Initial values

- $\lambda_0$ - the minimum value of $\lambda$,
- $\beta$ - the controlling variable determining the way the value of $\lambda$ changes with respect to the discounting factor $\delta$, i.e., boulware ($\beta < 1$), conceder ($\beta > 1$) or linear ($\beta = 1$),
- $\sigma^t$ - the estimation of the negotiating partner's concessive degree at time $t$,
- $\gamma$ - the controlling variable determining the way the value of $\lambda$ changes with respect to $\sigma^t$, i.e., boulware ($\gamma < 1$), conceder ($\gamma > 1$) or linear ($\gamma = 1$),
- $w$ - the weighting factor adjusting the relative effect of $\sigma^t$ on the nonexploitation point $\lambda$.

**if** $t = 0$ **then**
  $\lambda = \lambda_0 + (1 - \lambda_0)\delta^{\beta}$
**end if**
**if** $0 < t \leq 1$ **then**
  $\lambda = \lambda + w(1 - \lambda)\sigma^{t\gamma}$
**end if**

---

**Fig. 5.2** Adjustment rule of $\lambda$ at time $t$

(i.e., $\mathcal{C} = \{\omega \mid u_A(\omega) \geq l_A^t\}$). The negotiation outcome $\omega_{\max}$ will be returned by the NB component as the offer to be proposed to agent $B$.

To obtain $\omega_{\max}$, we need to estimate agent $B$'s private preference based on its past negotiation moves. Different approaches [3, 5, 7, 16] have been proposed to explicitly estimate the negotiating partner's utility function in bilateral negotiation scenarios. To make the estimation feasible with the limited information available, we usually need to put some restrictions on the possible structures that the negotiation partner's utility function can have [5] or assume that the preference profile of the negotiation partner is chosen from a fixed set of profiles [7]. Instead of estimating agent $B$'s utility function directly, here we adopt a model-free reinforcement learning-based approach to predict the current best negotiation outcome for agent $B$. The only assumption we need here is that the negotiating partner (agent $B$) is individually rational and follows some kind of concession-based strategy when proposing bids, which is the most commonly used assumption in both game-theoretic approaches and negotiations [5, 17].

Based on the above assumption, it is natural to assume that the sequence of past negotiation outcomes proposed by agent $B$ should be in accordance with the decreasing order of its preference over those outcomes. Intuitively, for a value $v_i$ of an item $m_i$, the earlier and the more frequent it appears in the negotiation outcomes of the past history, the more likely that it weights more in contributing to the negotiation partner's overall utility. Therefore, for each value of each item $m_i$ in the negotiation domain, we keep record of the number of times that it appears in the negotiating partner's past negotiation outcomes and update its value each time a new negotiation outcome $\omega'$ is proposed by agent $B$ as follows:

$$n(\omega'(m_i)) = n(\omega'(m_i)) + \eta^k \ \forall m_i \in \mathcal{M} \tag{5.4}$$

where $\omega'$ is the most recent negotiation outcome proposed by agent $B$, $\eta$ is the discounting factor reflecting the decreasing speed of the relative importance of the negotiation outcomes as time increases, and $k$ is the number of times that the value $\omega'(m_i)$ of item $m_i$ has appeared in the history.

For each negotiation outcome $\omega$, we define its accumulated frequency $f(\omega)$ as the criterion for evaluating the relative preference of agent $B$ over it. The value of $f(\omega)$ is determined by the value of $n(\omega(m_i))$ for each item $m_i \in \mathcal{M}$ based on the current negotiation history. Formally, for any negotiation outcome $\omega$, its accumulated frequency $f(\omega)$ is calculated as follows:

$$f(\omega) = \sum_{m_i} n(\omega(m_i)) \; \forall m_i \in \mathcal{M}. \tag{5.5}$$

The negotiation outcome $\omega_{\max}$ is selected based on the $\epsilon$-greedy exploration mechanism. With probability $1 - \epsilon$, it chooses the negotiation outcome with the highest $f$-value from the set $\mathcal{C}$ of candidate negotiation outcomes and chooses one negotiation outcome randomly from $\mathcal{C}$ with probability $\epsilon$. The value of $\epsilon$ controls the exploration degree during prediction. One exception is that the negotiation outcome $\omega_{\max}$ will be selected as the best negotiation outcome proposed by the negotiating agent $B$ in the history if the NB component has received the corresponding signal from the AC component, and this will be explained in details in next section.

### 5.3.3 Acceptance-Condition (AC) Component

Given the current negotiation history $H^t_{A\leftrightarrow B}$, agent $A$'s acceptance threshold $l^t_A$, and its negotiation outcome $x^t_{A\to B}$ to propose at time $t$, the AC component determines whether to accept the current proposal of agent $B$ or not. The overall acceptance conditions are described in Algorithm 5. The *ABiNeS* agent accepts the proposal $\omega_1$ from its negotiating agent $B$ if its utility over $\omega_1$ is either higher than its current acceptance threshold $l^t_A$ (Lines 2–3). Otherwise, it checks whether there exists some negotiation outcome $\omega_{\text{best}}$ previously proposed by its negotiating agent $B$ satisfying

**Algorithm 5** Acceptance conditions $AC(H^t_{A\leftrightarrow B}, l^t_A, x^t_{A\to B})$

1: Initialization: $\omega_{best}$ = BestNegotiationOutcome($H^t_{A\leftrightarrow B}$), and let $\omega_1$ be the negotiation outcome proposed by agent B obtained from $H^t_{A\leftrightarrow B}$.
2: **if** $u_A(\omega_1) > l^t_A$ **then**
3:     accept the offer (return true).
4: **else if** $u_A(\omega_{best}) > l^t_A$ **then**
5:     not accept (return false) and notify the NB component to propose the negotiation outcome $\omega_{best}$ next time.
6: **else**
7:     not accept (return false).
8: **end if**

**Algorithm 6** Termination conditions $TC(H^t_{A\leftrightarrow B}, l^t_A, x^t_{A\rightarrow B}, ru^0_A)$

1: **if** $ru^0_A > l^t_A$ **then**
2: accept the offer (return true).
3: **else**
4: not accept (return false).
5: **end if**

the above condition. If the answer is yes, then it will notify the NB component to propose $\omega_{\text{best}}$ next time (Lines 4–5).

### 5.3.4 *Termination-Condition (TC) Component*

This component is responsible for deciding whether to terminate the negotiation and receive the corresponding reservation value or not. Here we treat the reservation value as an alternative offer from the negotiating agent $B$ with a constant utility $ru^0_A$. Thus the termination conditions of TC component are similar to the acceptance conditions except that $u_A(\omega_1)$ is replaced with the reservation value $ru^0_A$. The only difference is that we do not need to check the best negotiation outcome proposed by agent $B$ in the history since the reservation value is unchanged throughout the negotiation session. The detailed mechanism for TC component is shown in Algorithm 6.

## 5.4 Experimental Simulations and Evaluations

In this section, we evaluate the negotiation performance of the *ABiNeS* strategy against the state-of-the-art negotiation agents across a wide range of multi-issue negotiation scenarios under the GENIUS (General Environment for Negotiation with Intelligent multipurpose Usage Simulation) [14] platform. GENIUS is a negotiation platform of ANAC [10, 11, 13] developed for facilitating research on bilateral multi-issue negotiations and allowing different negotiation agents to be evaluated in practical environments. All the requirements described in the negotiation model in Sect. 5.2 are supported in GENIUS.

Following the evaluation criterion in ANAC [10, 13, 18], first we evaluate the negotiation performance of the *ABiNeS* strategy against a number of state-of-the-art negotiation agents in the tournament setting. The overall performance of each negotiation strategy is calculated as its average utility obtained against all negotiation opponents averaged over all evaluation domains. We will first introduce the detailed experimental settings in Sect. 5.4.1, and the detailed evaluation results will be given in Sect. 5.4.2.

The evaluation criterion adopted in ANAC only reflects how a strategy performs within a particular negotiation setting; however, it does not reveal any information about how robust the strategy is if we allow the agents to change their strategies during negotiations. To this end, next we analyze the *robustness* of our strategy using the technique of empirical game theory [19] in Sect. 5.4.4. This technique has been successfully used to analyze the strategies in the past trading agent competitions [19] and also the robustness of the top eight strategies in ANAC 2011, which has been shown to be able to provide useful insights into different strategies. Following previous work [11, 20], we will apply this technique to analyze the negotiation results between the *ABiNeS* strategy and other top eight state-of-the-art negotiation strategies proposed in ANAC 2012 under different negotiation contexts.

## *5.4.1 Experimental Settings*

### 5.4.1.1 Negotiation Agents

The *ABiNeS* strategy is evaluated against the state-of-the-art negotiation strategies participated in ANAC 2012 [13]. The *ABiNeS* strategy is implemented as *CUHK-Agent* and has participated the ANAC 2012 competition. Therefore here we simply evaluate its performance[4] against another seven state-of-the-art negotiation agents in the final round of ANAC 2012 [13]: *AgentLG*, *OMACAgent* [21], *TheNegotiatior Reloaded*, *BRAMAgent2*, *Meta-Agent*, *IAMHaggler2012* [20, 22], and *AgentMR*. These negotiation agents are designed by different research groups independently and each agent has no information about other agents' strategies beforehand. Each agent can only learn about other agents' information within a single negotiation encounter, and all information learned at previous negotiation encounter will be erased at the beginning of the next encounter by the GENIUS platform. The detailed implementations of all negotiation strategies are available at [13].

### 5.4.1.2 Domains

The negotiation domains are designed by different research groups and are targeted at modeling practical multi-issue negotiation scenarios in uncertain and open environments. The domains are different from each other in terms of the number of issues, opposition [11], discounting factor, reservation values, and so on. Opposition is used for reflecting the competitiveness degree of the negotiation domain between the negotiation parties, which is defined as the minimum distance

[4]Here we only analyze the negotiation performance among these top eight negotiation strategies in the final round, but it is worth noticing that our *CUHKAgent* also ranks the 1st place in the qualifying round against a larger number of negotiation opponents (17 different teams).

from all the negotiation outcomes to the point representing complete satisfaction of both negotiation parties (1, 1). During negotiations, each agent can only have access to its own preference over all possible negotiation outcomes and has no information about its opponent's preference.

The basic set of the negotiation scenarios consists of 17 different domains submitted by the participants this year and also five domains from ANAC 2011 and two domains from ANAC 2010. For each negotiation domain from the basic set, three variants are also generated with different discounting factors and reservation values, which thus results in a total of 72 domains [13]. Next we use the *laptop* domain as an example to illustrate the characteristics as a negotiation domain. All other domains are similar and can be found at [13]. The outcome space of the *laptop* domain is shown graphically in Fig. 5.3. In this domain, a seller and a buyer negotiate over the specifications of a laptop in terms of three issues: the laptop brand, the hard disk size, and the size of the external monitor. Each issue has three possible options, which thus results in the total of 27 possible negotiation outcomes (the domain size). In Fig. 5.3, each point represents one possible negotiation outcome and the horizontal and vertical axis values of each point correspond to the utilities of the buyer and the seller, respectively. The solid line in the graph represents the Pareto frontier consisting of all the possible Pareto-optimal negotiation outcomes. The discounting factor and reservation value can be set to different values within the range of [0, 1], which may greatly influence the performance of negotiation strategies.

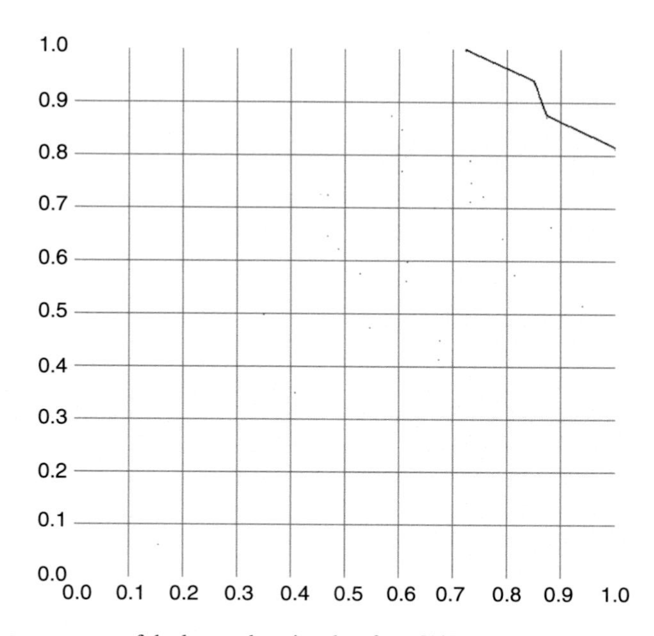


**Fig. 5.3** Outcome space of the laptop domain taken from [13]

**Table 5.1** Parameter setting of the *ABiNeS* strategy implemented as the *CUHKAgent* agent

| Parameters | Description | Values |
|---|---|---|
| $\alpha$ | Control the way the acceptance threshold is adjusted | 0 |
| $\lambda$ | The minimum value of nonexploitation point | 0.08 |
| $\beta$ | Control the way the nonexploitation point is adjusted with respect to the discounting factor of the domain | 1.2/1.5/1.8[a] |
| $\gamma$ | Control the way the nonexploitation point is adjusted with respect to the estimated concessive degree of the opponent | 10 |
| $\omega$ | Weighting factor adjusting the relative effect of the estimated concessive degree of the opponent on the value of the nonexploitation point | 0.1 |
| $\eta$ | The discounting factor reflecting the decreasing speed of the relative importance of the future negotiation outcomes | 1 |

[a]This parameter is set to different values depending on the value of the discounting factor $\delta$ of the current domain. If $\delta > 0.75$, then $\lambda = 1.8$; if $0.75 > \delta > 0.5$, then $\lambda = 1.5$; if $0 < \delta < 0.5$, then $\lambda = 1.2$

#### 5.4.1.3 Parameter Setting of the *ABiNeS* Strategy

The parameter setting of the *ABiNeS* strategy is given in Table 5.1, which follows the setting used in the *CUHKAgent* implementation.

### 5.4.2 Experimental Results and Analysis: Efficiency

In this section, we present the evaluation results over all domains, discounted domains and undiscounted domains. For each negotiation domain, each negotiation agent negotiates with all other seven agents (excluding itself) for 10 times on both sides in order to produce statistically significant results. All the results described in this section are adopted from the official final round results of ANAC 2012 [23].

#### 5.4.2.1 Performance over All Domains

Table 5.2 shows the overall ranking of the eight negotiation agents negotiating in terms of overall received utility averaged over all the 72 domains. We can see that our agent *ABiNeS* achieves the highest average score of 0.626478 and ranks the 1st place. *AgentLG* ranks the 2nd place with the average score of 0.621981, which is slightly lower than that of the *ABiNeS* agent. Both the *OMACAgent* and *TheNegotiator Reloaded* rank the 3rd place since there is no statistically significant difference between their scores.

**Table 5.2** Overall tournament scores of the negotiation agents average over all domains

| Agent name | Average score | Variance |
|---|---|---|
| *ABiNeS (CUHKAgent)* | 0.626478 | 0.000003 |
| *AgentLG* | 0.621981 | 0.000003 |
| *OMACAgent* | 0.618437 | 0.000002 |
| *TheNegotiator Reloaded* | 0.617247 | 0.000002 |
| *BRAMAgent2* | 0.592966 | 0.000002 |
| *Meta-Agent* | 0.586421 | 0.000003 |
| *IAMHaggler2012* | 0.535123 | 0.000001 |
| *AgentMR* | 0.328302 | 0.000003 |

**Table 5.3** Overall tournament scores of the negotiation agents average over all discounted domains

| Agent name | Average score | Variance |
|---|---|---|
| *ABiNeS (CUHKAgent)* | 0.577 | 0.000001 |
| *AgentLG* | 0.574 | 0.000002 |
| *OMACAgent* | 0.566 | 0.000001 |
| *BRAMAgent2* | 0.565 | 0.000003 |
| *TheNegotiator Reloaded* | 0.555 | 0.000002 |
| *Meta-Agent* | 0.551 | 0.000005 |
| *IAMHaggler2012* | 0.53 | 0.000001 |
| *AgentMR* | 0.361 | 0.000005 |

#### 5.4.2.2 Performance over Discounted and Undiscounted Domains

When the discounting effect is considered into the negotiation process, the agents need to make careful trade-off between exploiting and concession to the opponent. For the same domain, the negotiation outcomes when the discounting factor is considered are usually significantly different than those cases with no discounting effects. To this end, we further evaluate the negotiation performance of our *ABiNeS* agent against others under discounted and undiscounted domains.

Table 5.3 shows the scores of each negotiation agent averaged over all discounted domains. We can see that our *ABiNeS* agent achieves the highest average utility among all agents. We hypothesize that it is because our *ABiNeS* agent can adapt its negotiation decision more efficiently with respect to different discounting factors in terms of balancing between exploiting and concession to the opponent compared with others. Besides, it is worth noticing that the ranks of the *BRAMAgent2* and *TheNegotiator Reloaded* are reversed compared with the previous results averaged over all domains. This may indicate that *BRAMAgent2* agent can make better adaptation to different discounting factors than *TheNegotiator Reloaded* agent under discounted domains.

Next we consider the case of negotiation over undiscounted domains and the scores of all agents averaged over all undiscounted domains are shown in Table 5.4. From Table 5.4, we can see that there are significant changes among the relative ranking of the agents. *TheNegotiator Reloaded* agent ranks the 1st place, while our *ABiNeS* agent ranks the 2nd place. Under undiscounted domains, it is much easier

**Table 5.4** Overall tournament scores of the negotiation agents average over all undiscounted domains

| Agent name | Average score | Variance |
|---|---|---|
| *TheNegotiator Reloaded* | 0.742 | 0.000007 |
| *ABiNeS (CUHKAgent)* | 0.725 | 0.000012 |
| *OMACAgent* | 0.724 | 0.000006 |
| *AgentLG* | 0.717 | 0.000015 |
| *Meta-Agent* | 0.657 | 0.000004 |
| *BRAMAgent2* | 0.648 | 0.000009 |
| *IAMHaggler2012* | 0.546 | 0.000003 |
| *AgentMR* | 0.264 | 0.0 |

for agents to make decisions since there is no discounting effect on their actual received utilities as time passes. It is usually expected the less concessive an agent is, the more utility it will obtain under undiscounted domains. Accordingly, the overall ranking of the agents under undiscounted domains can largely reflect their relative concession degrees.

### 5.4.3 *Detailed Analysis of* ABiNeS *Strategy*

Previous section has shown the negotiation power of *ABiNeS* strategy against other state-of-the-art strategies under different negotiation domains. However, it is still unclear which decision-making component contributes to the superior negotiation efficiency of the *ABiNeS* strategy. Having a better understanding of this question can not only enable us to make further improvement of our strategy but also make it possible for other researchers to incorporate the decision-making components of our strategy into their negotiation strategy design appropriately.

In this section, we mainly focus on investigating two components inside our strategy as follows, which we believe make the most contributions to the overall negotiation performance of the *ABiNeS* strategy:

- the adaptive adjustment of the nonexploitation point $\lambda$ based on the behavior of the negotiating partner.
- the two-stage way in which the acceptance threshold $l_A^t$ is determined

During negotiations, on one hand, each rational agent would like to obtain as much payoff as possible by exploiting its negotiation partners; on the other hand, due to the discounting effect and time constraints, it is also necessary to make certain compromise to the negotiation partners to avoid obtaining very low utilities in the end (possibly caused by a break-off or high discounting effect). The concession degree should depend on both the characteristic of the negotiation domain and the type of the partner we are negotiating with. Therefore, one key problem during negotiation is how to balance the trade-off between exploiting and making compromise to the negotiating partner. In *ABiNeS*, the adaptive nonexploitation point $\lambda$ is introduced which represents the specific time when we should stop exploitations on the negotiating partner. The value of $\lambda$ is adaptively adjusted based on the estimated behavior of the negotiating partner (i.e., its concession degree).

We hypothesize that the adaptive adjustment of $\lambda$ can help to better exploit different negotiation opponents and thus improve the overall utility obtained through negotiation. To evaluate its influence on the overall negotiation performance of *ABiNeS* strategy, we compare the performance of the original *ABiNeS* strategy with the modified *ABiNeS* strategy in which the adaptive adjustment of $\lambda$ is removed (denoted as *ABiNeS*′). Table 5.5 shows the average utilities obtained by both strategies against other seven state-of-the-art strategies in ANAC 2012 in different domains. To make the evaluation results more general, for each domain the simulations are performed over four instances of the domains with different discounting factors ($\delta = 1, 0.75, 0.5, 0.25$). All results are average over 100 times. From Table 5.5, we can see that *ABiNeS* strategy can always obtain a higher average utility than *ABiNeS*′ in all negotiation domains. The results thus successfully verify our hypothesis of the usefulness of adaptively adjusting the value of $\lambda$ based on the estimated behavior of the opponent.

Next we evaluate the influence of the second component, the two-stage way in which the acceptance threshold $l_A^t$ is determined, on the negotiation efficiency of the *ABiNeS* strategy. The two-stage determination of the acceptance threshold

**Table 5.5** Comparison of the average tournament scores of *ABiNeS* an *ABiNeS*′ in different negotiation domains

| Domain name | *ABiNeS* (CUHKAgent) | *ABiNeS*′ |
|---|---|---|
| *Flight Booking* | 0.593165272 | 0.544166163 |
| *Supermarket* | 0.531929334 | 0.527879116 |
| *England vs Zimbabwe* | 0.642272791 | 0.641575379 |
| *Travel* | 0.624391428 | 0.620292902 |
| *Outfit* | 0.637956858 | 0.633111545 |
| *Grocery* | 0.669997176 | 0.66421635 |
| *Phone* | 0.681115608 | 0.674884384 |
| *Music Collection* | 0.737541642 | 0.731199686 |
| *Laptop* | 0.737647733 | 0.730444302 |
| *IS BT Acquisition* | 0.790048228 | 0.780533875 |
| *Camera* | 0.672763642 | 0.658183578 |
| *Housekeeping* | 0.617283604 | 0.602394404 |
| *Energy (small)* | 0.501314241 | 0.485674969 |
| *Itex vs Cypress* | 0.470921795 | 0.451796433 |
| *Energy* | 0.399831811 | 0.379083373 |
| *Amsterdam Party* | 0.673646911 | 0.652526336 |
| *Barbecue* | 0.674166041 | 0.63453876 |
| *Airport Site Selection* | 0.582970843 | 0.541472083 |
| *Barter* | 0.361420434 | 0.314742612 |
| *Rental House* | 0.526327346 | 0.479128338 |

works as follows: in the first stage ($t \leq \lambda$), the *ABiNeS* agent always makes gradual concession to its opponent following certain pattern of behavior but always maintain its acceptance threshold higher than the predetermined value $u^{\max}\delta^{1-t}$; in the second stage ($t > \lambda$), it accepts any proposal in which its discounted utility is no less than $u^{\max}\delta$. The rationale behind is that we assume that the opponents are always rational in that any positive proposal would be accepted by the opponent at the very last moment of the negotiation; therefore the worst case is that *ABiNeS* agent obtains the utility of $u^{\max}\delta$, i.e., proposing $u^{\max}$ at the very last moment of the negotiation to its opponent, and the proposal is accepted.

We hypothesize that this assumption is essential in guaranteeing that the *ABiNeS* agent would not be exploited significantly by its opponent during negotiation, and it is expected that the minimum utility $u^{\max}\delta$ can always be obtained no matter how aggressive the opponent is. To evaluate the influence of this assumption on the negotiation efficiency of the *ABiNeS* strategy, we compare the average utility of the original *ABiNeS* strategy against the state-of-the-art strategies with the modified version in which this assumption is removed (denoted as *ABiNeS''*). Table 5.6 shows the expected utilities obtained by both strategies against other seven state-of-the-art strategies in ANAC 2012 over different domains. Similar with the setting of previous part, for each domain we consider four instances of the domains with different

**Table 5.6** Comparison of the average tournament scores of *ABiNeS* and *ABiNeS''* in different negotiation domains

| Domain name | *ABiNeS* (CUHKAgent) | *ABiNeS''* |
|---|---|---|
| *Flight Booking* | 0.595633184 | 0.576213201 |
| *Supermarket* | 0.549546851 | 0.526440695 |
| *England vs Zimbabwe* | 0.708691163 | 0.586091569 |
| *Travel* | 0.625476427 | 0.5893065 |
| *Outfit* | 0.6371233 | 0.599609804 |
| *Grocery* | 0.635214787 | 0.634578829 |
| *Phone* | 0.670055909 | 0.623579617 |
| *Music Collection* | 0.74688424 | 0.716320117 |
| *Laptop* | 0.749388893 | 0.73743424 |
| *IS BT Acquisition* | 0.774180093 | 0.77663871 |
| *Camera* | 0.67820488 | 0.6581175 |
| *Housekeeping* | 0.621411774 | 0.602394404 |
| *Energy (Small)* | 0.483003949 | 0.471284284 |
| *Itex vs Cypress* | 0.470921795 | 0.481204633 |
| *Energy* | 0.399831811 | 0.379083373 |
| *Amsterdam Party* | 0.668931074 | 0.647760278 |
| *Barbecue* | 0.650610379 | 0.642122672 |
| *Airport Site Selection* | 0.569461523 | 0.560341328 |
| *Barter* | 0.465825872 | 0.463760361 |
| *Rental House* | 0.465322491 | 0.46809373 |

discounting factors ($\delta = 1, 0.75, 0.5, 0.25$). All results are average over 100 times. From Table 5.6, we can clearly see that *ABiNeS* agent is able to obtain higher average utility than *ABiNeS''* agent in all negotiation domains, which thus verifies the importance of adopting this assumption in improving the overall negotiation efficiency.

### 5.4.4 The Empirical Game-Theoretic Analysis: Robustness

The evaluation criterion adopted in previous sections reflects the negotiation performance of different strategies in terms of average scores achieved within a fixed tournament setting. However, it does not reveal much information about the *robustness* of the negotiation strategies in different negotiation scenarios, since it assumes that each agent's strategy is fixed beforehand. For example, we may be interested to know whether an agent adopting our *ABiNeS* strategy has the incentive to unilaterally deviate to other strategies under a particular negotiation tournament. Would any agent adopting other strategies be willing to switch to our strategy under different tournaments?

To answer the above question, we adopt the game-theoretic approach to analyze the robustness of our strategy under different negotiation settings. Since there exist an infinite number of possible negotiation strategies that the agents may take, we cannot apply the standard game-theoretic approach to perform such an analysis by explicitly considering all possible strategies. Therefore, here we adopt the tool of empirical game-theoretic (EGT) analysis to achieve this goal instead, which is originally developed by [19] to analyze the trading agent competition. EGT analysis is a game-theoretic analysis approach based on a set of empirical results. It handles the problem of the existence of infinite possible strategies by assuming that each agent only selects its strategy from a fixed set of strategies, and the outcomes for each strategy profile can be determined through empirical simulations. This technique has been successfully applied in addressing questions about robustness of different strategies from various domains including continuous double auction [24], trading strategies in previous years' TAC competitions [19], and negotiation strategies in ANAC 2011 [11]. Following previous work [11, 20], we apply the EGT analysis to the bilateral negotiation setting as follows.

In our EGT analysis, we consider a fixed set $\mathcal{S} = \{\mathbf{C}, \mathbf{L}, \mathbf{O}, \mathbf{R}, \mathbf{B}, \mathbf{M}, \mathbf{I}, \mathbf{A}\}$[5] of negotiation strategies consisting of the top eight strategies from this year's ANAC. Different from the setting of ANAC, each agent is free to select any strategy from this set as its negotiation strategy. For each bilateral negotiation, the corresponding payoff received for each participating agent is determined as its average payoff over

---

[5]The bold letters are the abbreviations for each strategy as follows: **C**, CUHKAgent; **L**, AgentLG; **O**, OMACAgent; **R**, TheNegotiatorReloaded; **B**, BRAMAgent2; **M**, Meta-agent; **I**, IAMHaggler2012; and **A**, AgentMR. These abbreviations will be used in the following descriptions.

**Table 5.7** Payoff matrix for the top eight negotiation strategies in ANAC 2012 average over all domains (For each strategy profile, only the row player's payoff is given since the game is symmetric.) The letters in bold are the abbreviations for each strategy

| Payoff matrix | **C** | **L** | **O** | **R** | **B** | **M** | **I** | **A** |
|---|---|---|---|---|---|---|---|---|
| **C** | 0.5956 | 0.465 | 0.491 | 0.669 | 0.548 | 0.618 | 0.832 | 0.437 |
| **L** | 0.541 | 0.4212 | 0.439 | 0.673 | 0.462 | 0.640 | 0.832 | 0 |
| **O** | 0.533 | 0.38 | 0.4233 | 0.648 | 0.433 | 0.562 | 0.815 | 0 |
| **R** | 0.546 | 0.522 | 0.502 | 0.5757 | 0.509 | 0.596 | 0.773 | 0.425 |
| **B** | 0.523 | 0.357 | 0.414 | 0.657 | 0.4626 | 0.648 | 0.757 | 0.207 |
| **M** | 0.501 | 0.486 | 0.472 | 0.623 | 0.484 | 0.4556 | 0.76 | 0.079 |
| **I** | 0.559 | 0.567 | 0.55 | 0.578 | 0.531 | 0.592 | 0.8192 | 0 |
| **A** | 0.471 | 0 | 0 | 0.615 | 0.163 | 0.12 | 0 | 0 |

all possible domains against its opponent, which can be obtained through empirical simulations. The detailed payoff matrix for all possible bilateral negotiations is given in Table 5.7. Note that for each strategy profile, only the row player's payoff is given since the game itself is agent-symmetric. Based on the bilateral negotiation outcomes in Table 5.7, it is easy for us to construct the negotiation outcomes (the corresponding payoff profiles) for any possible negotiation tournament among multiple agents. The average payoff of an agent in any given tournament can be determined by averaging its payoff obtained in all bilateral negotiations against all other agents participated in the tournament. Specifically, for a given tournament involving a set $\mathcal{P}$ of agents, the payoff $U_p(\mathcal{P})$ obtained by agent $p$ can be calculated as follows:

$$U_p(\mathcal{P}) = \frac{\sum_{p' \in \mathcal{P}, p' \neq p} U_p(p, p')}{|\mathcal{P}| - 1} \tag{5.6}$$

where $U_p(p, p')$ represents the corresponding average payoff of agent $p$ negotiating against another agent $p'$. Note that agents $p$ and $p'$ can use either the same or different strategies.

Given a negotiation tournament consisting of $n$ agents, each agent chooses one particular strategy from the set $\mathcal{S}$ of strategies, which jointly constitutes a strategy profile. Based on Eq. 5.6, now we can easily determine the corresponding payoff profile under this tournament. An agent has the incentive to deviate its current strategy to another one if and only if its payoff after deviation can be statistically significantly improved, provided that all the other agents keep their strategies unchanged. There may exist multiple candidate strategies that an agent has the incentive to deviate to, but here we only consider the best deviation available to that agent in terms of maximizing its deviation benefit following previous work [11]. Given a strategy profile under a negotiation tournament, if no agent has the incentive to unilaterally deviate from its current strategy, then this strategy profile is called an *empirical pure strategy Nash equilibrium*.[6] It is also possible for the agents to adopt mixed strategies and thus we can easily define the concept of *empirical mixed strategy Nash equilibrium* accordingly. However, in practical negotiations people are risk-averse and usually would like to be represented by a strategy with predictable behaviors instead of a probabilistic one [25, 26]. Therefore we only consider *empirical pure strategy Nash equilibria* in our analysis following previous work [11]. In general, a game may have no empirical pure strategy Nash equilibrium. Another useful concept for analyzing the stability of the strategy profiles is *best reply cycle* [27], which is a subset of strategy profiles in which, for any strategy profile within this subset, there is no single-agent best deviation path leading to any profile outside the cycle. In other words, in a best reply cycle, all single-agent best deviation paths starting from any strategy profile within itself must lead to another strategy profile inside the cycle.

Both *empirical pure strategy Nash equilibrium* and *best reply cycle* can be considered as two different interpretations of empirical stable sets to evaluate the *stability* of different strategy profiles. Based on these two concepts, we are ready to evaluate the *robustness* of a strategy using the concept of *basin of attraction* of a stable set [24]. The *basin of attraction* of a stable set is defined as the percentage of strategy profiles which can lead to this stable set through a series of single-agent best deviations. Accordingly, a negotiation strategy $s$ is considered to be *robust* if it belongs to a stable set with a large *basin of attraction* [11, 24]. In other words, if there exists a large proportion of initial strategy profiles which, through a series of single-agent best deviations, can eventually lead to a stable set containing strategy $s$, then strategy $s$ is highly robust in the long run, since the strategy $s$ can always have the opportunity of being adopted eventually if the tournament is sufficiently repeated following single-agent best deviations.

We apply the EGT analysis to identify both *empirical pure strategy Nash equilibrium* and *best reply cycle* and evaluate the *robustness* of our *ABiNeS* strategy based on the *basin of attraction* of the above two stable sets in the following two negotiation contexts:

- Bilateral negotiations in which each agent is allowed to choose any strategy from the set $S$
- Negotiation tournaments following the setting of ANAC 2012 competition, except that the agents are allowed to select the same strategy during the same tournament

[6]This concept is similar to the concept of *pure strategy Nash equilibrium* in classical game theory, but it is called *empirical pure strategy Nash equilibrium* since the analysis is based on empirical results.

#### 5.4.4.1 Applying Model Checking to Perform EGT Analysis

We propose applying model checking techniques [28] to perform EGT analysis by automatically identifying both *empirical pure strategy Nash equilibrium* and *best reply cycle* and also determining the *basin of attraction* of the above two stable sets. Model checking is a formal verification technique for analyzing the behaviors of deterministic or stochastic systems. Usually a mathematical model (e.g., Markov decision process (MDP)) of the system is constructed first and the corresponding properties we would like to check have to be formally specified. Then a quantitative analysis of the model is performed by applying a combination of exhaustive search and numerical solution methods with respect to the property to be checked.

The dynamics of the agents' negotiation strategy changes can be naturally modeled as an MDP in which each state is represented by the number of agents choosing each strategy, and the transition between each pair of states $(s_1, s_2)$ indicates there exists a single-agent best deviation from state $s_1$ to $s_2$. In this MDP, there may exist nondeterminism since each state may have multiple nondeterministic outgoing transitions to other states since different players may adopt different strategies and their best deviations may be different. However, each transition is deterministic since we only consider best deviation for each agent. The tasks of identifying both *empirical pure strategy Nash equilibrium* and *best reply cycle* and determining the *basin of attraction* of the above two stable sets can also be easily expressed or transformed as properties supported by current model checkers. Therefore it is natural for us to apply model checking techniques to perform the EGT analysis.

We adopt the model checker PAT [29] to perform the EGT analysis, which is able to automatically check all existing *empirical pure strategy Nash equilibrium* and *best reply cycle* and also determining their *basin of attraction* for different negotiation contexts. Besides, another advantage of using PAT is that it can visualize the transition graph among all possible states, which can help us to have a better understanding of the dynamics of the agents' negotiation strategy changes from the global perspective. We omit the details of how to build the models using PAT and the its detailed implementation of checking those properties since this is not our focus here, and interested readers may refer to [30] for details.

Bilateral Negotiations Among Eight Possible Strategies

In the context of bilateral negotiations, there exist a set $\mathcal{P}$ of agents ( $|\mathcal{P}| = 2$) and a set $\mathcal{S} = \{\mathbf{C}, \mathbf{L}, \mathbf{O}, \mathbf{R}, \mathbf{B}, \mathbf{M}, \mathbf{I}, \mathbf{A}\}$ of strategies. Each agent can choose any strategy from the set $\mathcal{S}$ during negotiation. In general there exist a total of $|\mathcal{S}|^{|\mathcal{P}|} = 8^2 = 64$ possible strategy profiles. However, since the bilateral negotiation itself is symmetric (e.g., a negotiation in which agent $p_1$ uses strategy **C** and agent $p_2$ uses strategy **L** is equivalent with the case that their strategies are swapped), we can simply reduce the number of strategy profiles from 64 to $\binom{|\mathcal{P}|+|\mathcal{S}|-1}{|\mathcal{S}|-1} = 36$. This also indicates that the number of states of the MDP model generated by PAT for performing EGT analysis can be reduced from 64 to 36, which can help reducing the verification time of PAT.

According to the verification results of PAT, it is found out that under bilateral negotiations, there is no *empirical pure strategy Nash equilibrium* and there exists only one *best reply cycle*, i.e., $(L, C) \rightarrow (R, L) \rightarrow (C, R) \rightarrow (L, C)$. Besides, the *basin of attraction* of this cycle is 100 %, i.e., for all possible initial strategy profiles, there always exists a single-agent best deviation path which can lead to one of the strategy profiles within this cycle. Our *ABiNeS* strategy is contained in two strategy profiles ($(C, R)$ and $(C, L)$) in this cycle. This indicates that our *ABiNeS* strategy is very robust against other strategies since it is always possible that the agents will be willing to adopt our *ABiNeS* strategy no matter what their initial strategy is in the long run.

Figure 5.4 shows an example of the dynamics of how each agent may change its strategy between different negotiation tournaments under bilateral negotiations among eight strategies, and this graph is generated automatically by the model checker PAT. In Fig. 5.4, each node (state) is associated with a unique number representing a unique bilateral negotiation setting (strategy profile) except node 1 which, as the initial state, consists of all the possible bilateral negotiation settings

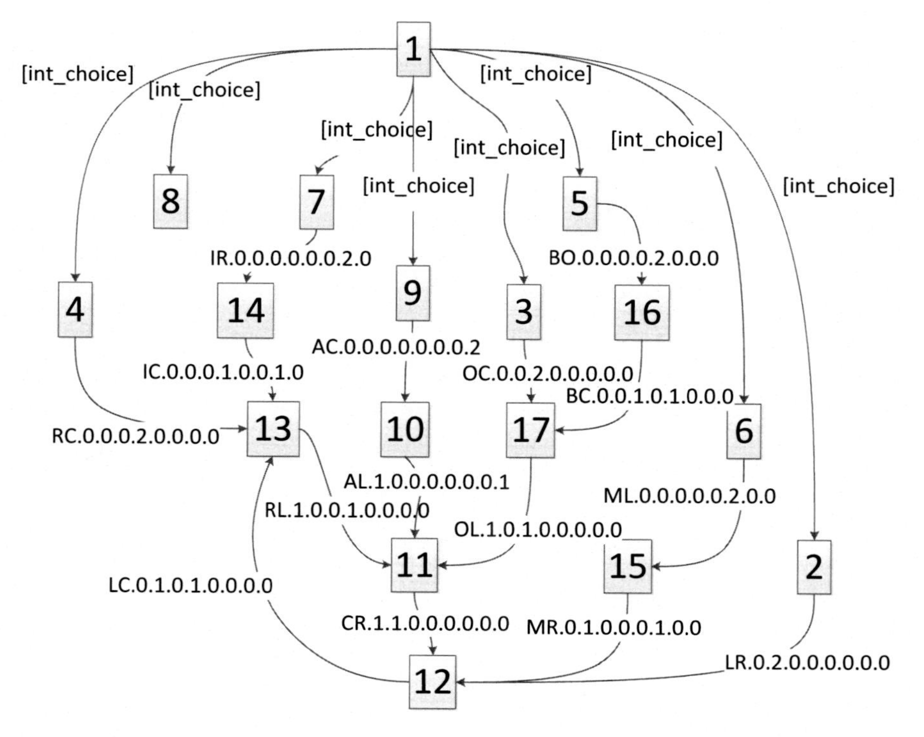


**Fig. 5.4** Deviation analysis with initial state (node 1) consisting of all possible bilateral negotiation settings in which both agents choose the same strategy from the eight possible strategies

in which both agents choose the same strategy. The strategy profile for each state is given by the label of its outgoing transitions. Each transition between any pair of nodes (states) indicates a single-agent best deviation for one particular type of strategy. For example, considering node 4 and node 13, node 4 represents the bilateral negotiation in which both agents use strategy **R** and node 13 represents the bilateral negotiation in which one agent chooses strategy **C**, while the other one chooses strategy **R**, and the transition between them indicates that there exists such a single-agent best deviation from node 4 to node 13. From this graph, we can also easily check that there only exists one *best reply cycle* from node 13 to node 11 to node 12 and back to node 13, and for any initial negotiation settings (strategy profiles) in the initial state (node 1), all of them will finally converge to one node within this cycle through a series of single-agent best deviations.

However, the analysis within bilateral negotiation setting does not give us much information about the robustness of our strategy within a tournament setting involving more than two agents, which will be described in next section.

### Eight-Agent Negotiation Tournaments

In this section, we analyze the robustness of our *ABiNeS* strategy within the context of the negotiation tournament among eight players. We start with a simple case in which each agent is only allowed to choose one strategy from the top four strategies,[7] and then we perform the robustness analysis by taking all top eight strategies into consideration following the setting of ANAC. For both cases, each participating agent negotiates with all the other participants, and the average payoff of each agent can be determined using Eq. 5.6. However, different from the setting in ANAC, the agents are free to choose any strategy from the set of strategies available and thus different agents may select the same strategy during the same tournament in our EGT analysis. The setting of ANAC can be considered as a specific tournament in which each agent chooses a unique strategy among the eight strategies.

One natural way is to perform robust analysis over all domains; however, this can hide a lot of detailed information due to the averaging effects. Besides, most of the domains are relatively small and thus more easy for the agents to negotiate to get a high utility under the limited negotiation time (3 min). Therefore, we conduct the robustness analysis under the tournament setting over the challenging domain: *Travel* domain, similar to previous work [20]. The *Travel* domain is one of the largest and most complex domains in the competition, which thus can better reflect the practical negotiation scenarios which usually involve a large number of possible proposals to consider. In addition, different from [20], we set the discounting factor of this domain to the low value of 0.5 instead of 1 (without discounting). Under the setting with high discounting effect, it requires the agents to delicately and

[7] The reason that we choose the top four strategies instead of the top three is that both *OMACagent* and *TheNegotiatorReloaded* rank the third place.

adaptively trade off between concession to the opponent (be fear of obtaining lower payoff due to large discounting effect) and staying tough (hope to get higher payoff by letting the opponent concede first) against different types of opponents. Therefore we believe that this setting can better reflect the behavior differences between different strategies.

First we consider the eight-player negotiation tournament over the set $\mathcal{S} = \{C, L, O, R\}$ of the top tour strategies. Similar with the analysis in bilateral negotiation, the total number of strategy profiles considered can be reduced from $|\mathcal{S}|^{|\mathcal{P}|} = 65{,}536$ to $\binom{|\mathcal{P}|+|\mathcal{S}|-1}{|\mathcal{S}|-1} = 165$ considering the symmetry of the negotiation.

The robustness analysis is performed using the model checker PAT [29]. Based on the verification from PAT, we observe that there only exists one *empirical pure strategy Nash equilibrium*, $(C, C, C, C, C, C, C, C)$, in which all agents adopt our *ABiNeS* strategy, and also for all nonequilibrium tournaments, there always exists a single-agent best deviation path leading to this equilibrium. This result indicates that our strategy is very robust under the setting of negotiation tournament, even though the agents' average payoff under this tournament is lower compared with some other tournaments (e.g., all agents adopting the strategy *IAMHaggler2012*[8]). Similar phenomena (the inefficiency of Nash equilibrium from the social perspective) are commonly observed in noncooperative game theory. For example, in the prisoner's dilemma game, mutual defection is the only pure strategy Nash equilibrium, but there exists another Pareto-optimal outcome of mutual cooperation under which all agents' payoffs can be significantly increased.

We give a specific illustration of how the agents adjust their strategy choices between different tournaments under the EGT analysis in Fig. 5.5. This deviation analysis graph is generated automatically using the model checker PAT's simulation tool. In Fig. 5.5, each node is associated with a unique number representing different tournaments. The strategy profile for each state is given by the label of its outgoing transitions. Each transition indicates a single-agent best deviation for one particular type of strategy. For example, node 2 represents the tournament in which the number of agents choosing the four strategies $C, L, O, R$ are 3, 1, 2, 2, respectively, which is specified by the labels of its outgoing transitions *LC*.**3.1.2.2**. The transition from node 2 to node 27 indicates that there exists a single-agent best deviation for an agent choosing strategy $L$ to deviate to strategy $C$, as indicated by its label **LC**.3.1.2.2. From Fig. 5.5, we can clearly observe the overall trend that all agents choosing strategies different from $C$ have the incentive to deviate from their current strategies to strategy $C$. This deviation analysis graph can be viewed as a graph with six levels, and the transitions between every adjacent levels $i$ and $j$ correspond to single-agent best deviations from states in level $i$ to states in level $j$. At the last level, all agents originally choosing strategies different from $C$ have switched to strategy $C$, thus converging to the terminating node 13 which corresponds to the pure strategy Nash equilibrium $(C, C, C, C, C, C, C, C)$.

[8]This strategy wins the *most social agent* award in ANAC 2012 since it achieves the highest social payoff (the sum of its own and its opponent's payoffs).

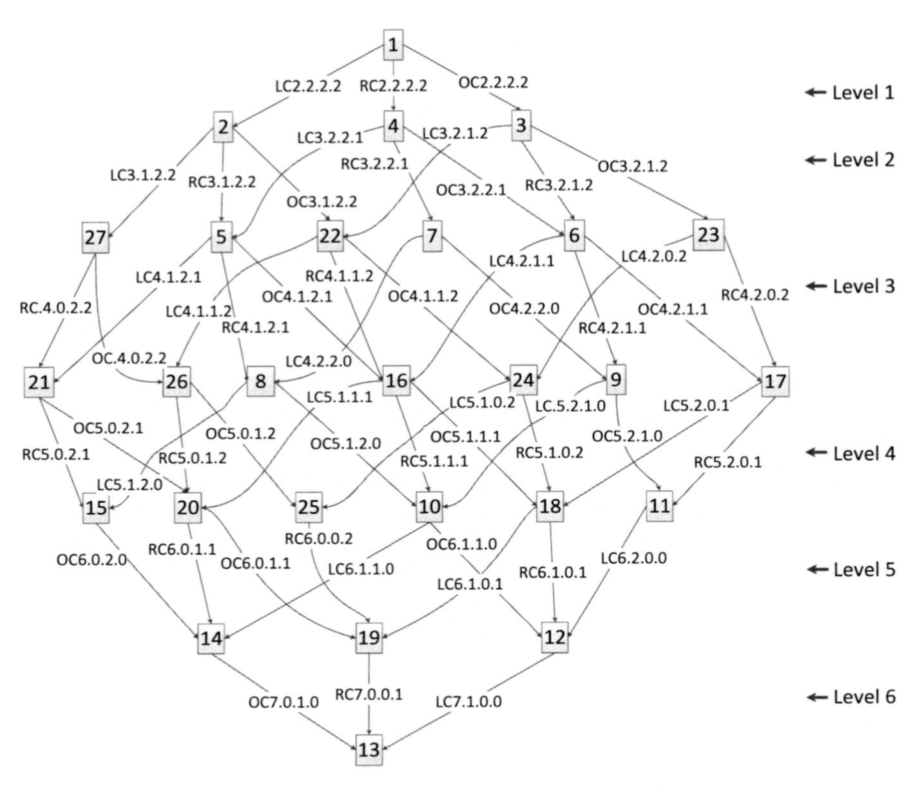


**Fig. 5.5** Deviation analysis with initial state (node 1) in which each strategy is chosen by two agents

Next we turn to the analysis of the complete eight-player tournament over the top eight strategies following the setting of ANAC 2012. In this setting, the total number of strategy profiles considered can be reduced from $|\mathcal{S}|^{|\mathcal{P}|} = 16{,}777{,}216$ to $\binom{|\mathcal{P}|+|\mathcal{S}|-1}{|\mathcal{S}|-1} = 6435$ due to the symmetry of the negotiation. Similar result as that in the previous case of the top four strategies can be observed. There exists only one *empirical pure strategy Nash equilibrium*, $(C, C, C, C, C, C, C, C)$, where all agents adopt our *ABiNeS* strategy, and also the *basin of attraction* of this equilibrium is 100 %. This indicates that our *ABiNeS* strategy is also robust under the full negotiation tournament setting over top eight strategies. Similar deviation trends as the case of the top four strategies can also be found here: for any initial tournament, those agents choosing strategies other than strategy $C$ have the incentive to switch to strategy $C$ to maximally increase their individual average payoffs, and thus after a series of single-agent best deviations, finally the strategy profile will converge to the pure strategy Nash equilibrium $(C, C, C, C, C, C, C, C)$.

## 5.5 Conclusion

In this chapter, we focus on a competitive bilateral negotiation environment in which two selfish agents negotiate with each other to maximize their own individual benefits. We aim at investigating the problem of how an individual agent can maximize its benefit from negotiation by exploiting its negotiating partners as much as possible. To this end, we describe an adaptive negotiation *ABiNeS* strategy which introduces the concept of nonexploitation point $\lambda$ to adaptively adjust the *ABiNeS* agent's concession degree to its negotiating opponent. Besides, a reinforcement learning-based approach is adopted to determine the optimal proposal for the negotiating partner to maximize the possibility that the offer will be accepted.

The performance of the *ABiNeS* strategy is evaluated using two different measures: *efficiency* (average payoff) within a single negotiation tournament and *robustness* which is determined by the size of the basin of attraction of those strategy profiles that our strategy belongs to under different negotiation tournament settings. The *ABiNeS* strategy is shown to be very efficient against the state-of-the-art strategies from ANAC 2012 and can obtain the highest average payoff over a large number of negotiation domains. Detailed analysis of the *ABiNeS* strategy in terms of the influences of its two major decision components on the negotiation efficiency is also provided, which gives us valuable insights of why it can win the championship in ANAC 2012. Last but not the least, we provide a novel way of applying model checking techniques to perform EGT analysis to determine the robustness of the strategies. The *ABiNeS* strategy is found to be very robust in both bilateral negotiations and negotiation tournaments among eight players following the setting of ANAC.

## References

1. Hao JY, Leung HF (2012) Abines: an adaptive bilateral negotiating strategy over multiple items. In: Proceedings of IAT'12, Macau, vol 2, pp 95–102
2. Lai GM, Li CH, Sycara K, Giampapa JA (2004) Literature review on multi-attribute negotiations. Technical report CMU-RI-TR-04-66, Robotics Institute, Carnegie Mellon University, Pittsburgh, Dec 2004
3. Faratin P, Sierra C, Jennings NR (2003) Using similarity criteria to make negotiation trade-offs. Artif Intell 142(2):205–237
4. Saha S, Biswas A, Sen S (2005) Modeling opponent decision in repeated one-shot negotiations. In: Proceedings of AAMAS'05, Utrecht, pp 397–403
5. Hindriks K, Tykhonov D (2008) Opponent modeling in auomated multi-issue negotiation using bayesian learning. In: Proceedings of AAMAS'08, Estoril, pp 331–338
6. Jakub B, Ryszard K (2006) Predicting partner's behaviour in agent negotiation. In: Proceedings of AAMAS'06, Hakodate, pp 355–361
7. Zeng D, Sycara K (1998) Bayesian learning in negotiation. Int J Hum Comput Syst 48:125–141
8. Coehoorn RM, Jennings NR (2004) Learning an opponent's preferences to make effective multi-issue negotiation trade-offs. In: Proceedings of ICEC'04, Delft, pp 59–68

9. Zeng D, Sycara K (1996) Bayesian learning in negotiation. In: AAAI symposium on adaptation, co-evolution and learning in multiagent systems, Portland, pp 99–104
10. Baarslag T, Hindriks K, Jonker C, Kraus S, Lin R (2010) The first automated negotiating agents competition (ANAC 2010). In: Ito T, Zhang M, Robu V, Fatima S, Matsuo T (eds) New trends in agent-based complex automated negotiations. Springer, Berlin/Heidelberg, pp 113–135
11. Baarslag T, Fujita K, Gerding EH, Hindriks K, Ito T, Jennings NR, Jonker C, Kraus S, Lin R, Robu V, Williams CR (2013) Evaluating practical negotiating agents: results and analysis of the 2011 international competition. Artif Intell 198:73–103
12. Baarslag T, Hindriks K, Jonker C (2011) Acceptance conditions in automated negotiation. In: Proceedings of ACAN'11, Taibei
13. The third international automated negotiating agent competition (ANAC 2012) (2012). http://anac2012.ecs.soton.ac.uk/. Accessed 30 Sept 2015
14. Lin R, Kraus S, Baarslag T, Tykhonov D, Hindriks K, Jonker CM (2014) Genius: an integrated environment for supporting the design of generic automated negotiators. Comput Intell 30(1):48–70
15. Song S, Hao J, Liu Y, Sun J, Leung H-F, Zhang J (2015) Improved EGT-based robustness analysis of negotiation strategies in multiagent systems via model checking. IEEE Trans Hum-Mach Syst
16. Coehoorn RM, Jennings NR (2004) Learning an opponent's preferences to make effective multi-issue negotiation trade-offs. In: Proceedings of ICEC'04, Delft, pp 59–68. ACM
17. Osborne MJ, Rubinstein A (1994) A course in game theory. MIT, Cambridge
18. The second international automated negotiating agent competition (ANAC 2011) (2011). http://www.itolab.nitech.ac.jp/ANAC2011/. Accessed 30 Sept 2015
19. Estelle J, Wellman MP, Singh S, Vorbeychik Y, Soni V (2005) Strategic interactions in a supply chain game. Comput Intell 21(1):1–26
20. Williams CR, Robu V, Gerding EH, Jennings NR (2012) Using Gaussian processes to optimise concession in complex negotiations against unknown opponents. In: Proceedings of IJCAI'12, Montpellier, pp 432–438
21. Chen SQ, Weiss G (2012) An efficient and adaptive approach to negotiation in complex environments. In: Proceedings of ECAI'12, Montpellier, pp 228–233
22. Williams CR, Robu V, Gerding EH, Jennings NR (2012) Negotiating concurrently with unkown opponents in complex, real-time domains. In: Proceedings of ECAI'12, Montpellier, pp 834–839
23. Results of the third international automated negotiating agent competition (ANAC 2012) (2012). http://anac2012.ecs.soton.ac.uk/results/final/. Accessed 30 Sept 2015
24. Vytelingum P, Cliff D, Jennings NR (2008) Strategic bidding in continuous double auctions. Artif Intell 172(14):1700–1729
25. Pratt JW (1964) Risk aversion in the small and in the large. Econometrica 32:122–136
26. Arrow KJ (1971) The theory of risk aversion. In: Arrow KJ (ed) Essays in the theory of risk-bearing. Markham Publishing Company, Chicago, pp 90–120
27. Yong H (1993) The evolution of conventions. Econometrica 61(1):57–84
28. Clarke EM, Grumberg O, Long DE (1994) Model checking and abstraction. ACM Trans Program Lang Syst 16(5):1512–1542
29. Sun J, Liu Y, Dong JS, Chen C (2009) Integrating specification and programs for system modeling and verification. In: Proceedings of TASE'09. IEEE Computer Society, Los Alamitos, pp 127–135
30. Sun J, Liu Y, Dong JS, Pang J (2009) PAT: towards flexible verification under fairness. In: Proceedings of CAV'09, Grenoble, pp 709–714

# Chapter 6
# Social Optimality in Competitive Multiagent Systems

In this chapter, we turn to investigate the question of how socially optimal solutions can be achieved in competitive MASs which consist of individually rational agents. In competitive multiagent environments, the individually rational agents are assumed to be only interested in maximizing their individual benefits and may not be willing to follow the (socially oriented) strategy specified by the system designer. If each agent behaves in a purely individually rational manner, it usually leads to the nonsocially optimal outcomes, thus decreasing the overall system's utilities. Therefore we usually need to resort to designing effective mechanisms to motivate those selfish agents to change their behaviors toward coordinating on socially optimal outcomes. Specifically in this chapter we look at how to handle this problem within two major learning frameworks in MASs. The first one is the infinitely repeated game learning framework [1, 2], which will be introduced in Sect. 6.1. The second one is the social learning framework which will be introduced in Sect. 6.2 [3].

## 6.1 Achieving Socially Optimal Solutions in the Context of Infinitely Repeated Games

One most commonly adopted interacting framework in multiagent learning literature is two-player repeated games, in which each agent chooses its action independently and simultaneously each round. The most commonly adopted assumption is that the agents are willing to follow the same learning strategy as designed. However, the weakness of this assumption is that we usually may not have control on the behaviors of all agents. Within an open environment, the agents are usually designed by different parties and may have not the incentive to follow the strategy we design. To this end, we adopt the "AI agenda" [4] within a "competitive" environment by assuming that the opponent agent will not adopt the strategy we



J. Hao, H.-f. Leung, *Interactions in Multiagent Systems: Fairness, Social Optimality and Individual Rationality*, DOI 10.1007/978-3-662-49470-7_6

design. We assume that the opponent agent is individually rational and may adopt one of the following well-known rational strategies: $Q$-learning [5], WoLF-PHC [6], and fictitious play (FP) [7] following previous work [8]. In "AI agenda" [4], one commonly pursued direction is considering how to obtain as high rewards as possible by exploiting the opponents [9]. However, we are more interested in how the opponents can be influenced toward coordination on socially optimal outcomes through repeated interactions [8] from the system designer's perspective.

### *6.1.1 Learning Environment and Goal*

We focus on the class of two-player repeated normal-form games. At the end of each round, each agent receives its own payoff based on the outcome and also observes the action of its opponent. Two examples of normal-form games (prisoner's dilemma game and Stackelberg game) are already shown in Figs. 6.1 and 6.2, respectively.

Following the setting in [8, 10], we assume that the opponent is individually rational, and specifically we consider the opponent may adopt one of the following well-known rational strategies: $Q$-learning [5], WoLF-PHC [6],[1] and fictitious play [7]. $Q$-learning is a rational learning algorithm that has been widely applied in multiagent interacting environments. It has been proved that the agents using $Q$-learning algorithm converge to some pure strategy Nash equilibrium in deterministic cooperative games only [11], but no guarantees on which Nash equilibrium that will be converged to. WoLF-PHC is empirically shown to converge to a Nash equilibrium in two-player two-action games; however, similar with $Q$-learning, the Nash equilibrium that the agents converge to may be extremely inefficient. Finally,

**Fig. 6.1** Payoff matrix for prisoner's dilemma game

| 1's payoff, 2's payoff | | Player 2's action | |
|---|---|---|---|
| | | C | D |
| Player 1's action | C | 3, 3 | 0, 5 |
| | D | 5, 0 | 1, 1 |

**Fig. 6.2** Payoff matrix for Stackelberg game

| 1's payoff, 2's payoff | | Player 2's action | |
|---|---|---|---|
| | | L | R |
| Player 1's action | U | 1, 0 | 3, 2 |
| | D | 2, 1 | 4, 0 |

[1] WoLF-PHC is short for Win or Learn Fast—policy hill climbing.

fictitious play is a rational learning strategy widely studied in game theory literature, and it is guaranteed to converge to a Nash equilibrium in certain restricted classes of games (e.g., games solvable by iterated elimination of strictly dominated strategies). Under fictitious play, each agent keeps the record of its opponent's action history and chooses its action to maximize its own expected payoff with respect to its opponent's mixed strategy (obtained from the empirical distribution of its past action choices). For our purpose, we do not take into consideration the task of learning the game itself and assume that the game structure is known to both agents beforehand.

It is well known that every two-player normal-form game with finite actions has at least one pure/mixed strategy Nash equilibrium [12]. Under a Nash equilibrium, each agent is making its best response to the strategy of the other agent and thus no agent has the incentive to unilaterally deviate from its current strategy. Nash equilibrium in single-stage games has been commonly adopted as the learning goal to pursue in previous work [5, 6, 13, 14]; however, it can be extremely inefficient in terms of the payoffs the agents receive (see the example in Fig. 6.1). To this end, we set our learning goal to converging to socially optimal outcome sustained by Nash equilibrium (SOSNE). A SOSNE is an outcome which is socially optimal in the single-stage game in terms of maximizing the sum of all players' payoffs and also corresponds to a Nash equilibrium payoff profile when the game is infinitely repeated with limit of means criterion.[2] Compared with converging to Nash equilibrium in single-stage games, converging to a SOSNE outcome not only achieves system-level efficiency but also maintains system stability since any agent deviating from the SOSNE outcome can be punished successfully which is guaranteed by the Nash folk theorem [12]. For example, consider the repeated prisoner's dilemma game in Fig. 6.1. $(C, C)$ is the only SOSNE outcome in this game, since it is the only outcome under which the sum of both agents' payoffs is maximized, and also it corresponds to a Nash equilibrium payoff profile in the limit of means infinitely repeated PD game.

**Definition 6.1** For any two-player normal-form game $G$, let us denote the set of Nash equilibrium payoff profiles in the corresponding infinitely repeated game under the limit of means criterion as $\mathcal{P}$. A $(s_1^*, s_2^*)$ is a *socially optimal outcome sustained by Nash equilibrium (SOSNE)* if and only if there exists a payoff profile $(p_1, p_2) \in \mathcal{P}$ such that $u_1(s_1^*, s_2^*) = p_1$ and $u_2(s_1^*, s_2^*) = p_2$.

From the Nash folk theorem [12], we know that, for any two-player game, any feasible payoff profile which Pareto-dominates the minimax payoff profile (or feasible enforceable payoff profile) corresponds to a Nash equilibrium of the limit of means infinitely repeated game. Accordingly, to check whether an outcome is a SOSNE outcome, we only need to examine whether it is socially optimal and

[2]A preference relation $\succsim_i$ for player $i$ is defined under the limit of means criterion if it satisfies the following property: $O_1 \succsim_i O_2$ if and only if $\lim_{t\to\infty} \Sigma_{k=1}^{t}(p_1^k - p_2^k)/t \geq 0$, where $O_1 = (a_{i,t}^1, a_{j,t}^1)_{t=1}^{\infty}$, and $O_2 = (a_{i,t}^2, a_{j,t}^2)_{t=1}^{\infty}$ are the outcomes of the infinitely repeated game and $p_1^k$ and $p_2^k$ are the corresponding payoffs player $i$ receives in round $k$ of outcomes $O_1$ and $O_2$, respectively.

also Pareto-dominates the minimax payoff profile of the single-stage game. Taking the Stackelberg game shown in Fig. 6.2 as an example, let us represent each payoff profile as a point in two-dimensional space shown in Fig. 6.3. In this figure, the x-axis represents the payoff to the row player (player 1) and the y-axis denotes the payoff to the column player (player 2). All the feasible payoff profiles in repeated game under the limit of means criterion are within the triangle area with three vertices of (1, 0), (4, 0), and (3, 2), and also it is easy to check that the minimax payoff profile corresponds to the point (2, 1). A payoff profile is Pareto-dominated by all feasible payoff profiles that lie in the right or above it. Therefore, in this example, the set $\mathcal{P}$ of Nash equilibrium payoff profiles (i.e., the set of points Pareto-dominating the minimax payoff profile (2, 1)) is represented by the subtriangle with the three vertices: minimax, $c$, and $SO$. The payoff profile $SO$ is socially optimal and also is within this subtriangle region; thus apparently it is a Nash equilibrium payoff profile in the limit of means infinitely repeated Stackelberg game. Therefore the outcome $(U, R)$ corresponding to the payoff profile $SO$ is a SOSNE outcome. If there exists a SOSNE outcome in a game, then it is reasonable to expect that any individually rational agent would have the incentive to coordinate on it given the threat of being punished by obtaining the worse minimax payoff otherwise. Notice that there may not exist a SOSNE outcome in certain games. For example, consider the game shown in Fig. 6.4. In this game, it is easy to check that only the outcome $(C, C)$ is a POSNE outcome but not a SOSNE outcome. In this case, we will adopt the socially optimal outcomes (e.g., $(D, D)$ in Fig. 6.4) as the target solutions to pursue.

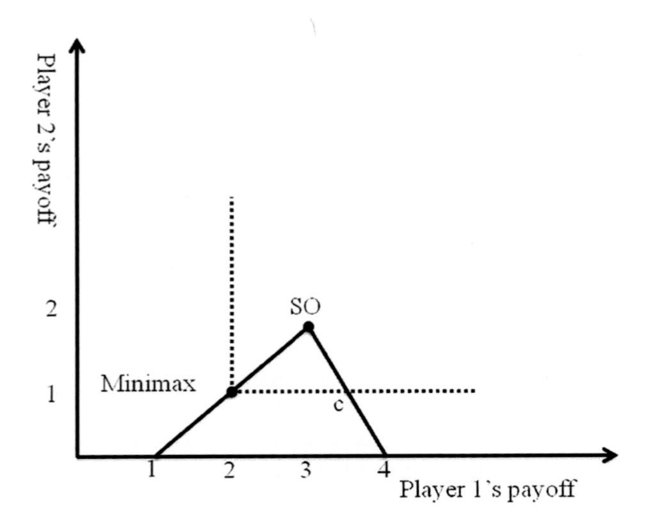


**Fig. 6.3** Payoff profile space of the Stackelberg game in Fig. 6.2

**Fig. 6.4** A two-player game without a SOSNE outcome

| 1's payoff, 2's payoff | | Player 2's action | |
|---|---|---|---|
| | | C | D |
| Player 1's action | C | 3, 2 | 4, 1 |
| | D | 1, 3 | 2, 4 |

## 6.1.2 *TaFSO: A Learning Approach Toward SOSNE Outcomes*

In this section, we present the learning approach TaFSO aiming at achieving SOSNE outcomes. To enable the TaFSO strategy to exert effective influence on the opponent's behavior, we consider an interesting variation of sequential play by allowing entrusting decision to others. During each round, apart from choosing an action from its original action space, every agent is also given an additional option of asking its opponent to make the decision for itself.

The strategy TaFSO combines the properties of both *oracle* and *follower* strategies based on the action entrustment mechanism. Its *oracle* component is used to influence its opponent to cooperate and behave in the expected way based on punishment and reward mechanisms, which mainly involves the following two functions:

- If the opponent agent chooses action $F$, the *oracle* component will be responsible for determining which joint action to execute to reward the opponent;
- Otherwise the *oracle* component will determine the suitable set of actions for punishing the opponent for being uncooperative.

Its *follower* component is in charge of which action to choose to punish the opponent agent while ensuring that the TaFSO agent can obtain as much payoff as possible against its opponent at the same time.

### 6.1.2.1 *Oracle* Strategy in TaFSO

During each round, apart from choosing an action from its original action space, every agent is also given an additional option of asking its opponent to make the decision for itself. Whenever the opponent $j$ decides to entrust TaFSO agent $i$ to make decisions (denoted as choosing action $F$), TaFSO agent $i$ will select the socially optimal joint action pair $(s_1, s_2)$ and both agents will execute their corresponding actions accordingly. If there exist multiple action pairs that are socially optimal, then the one with higher payoff for the opponent agent is selected. We assume that every agent will honestly execute the action assigned by its opponent whenever it asks its opponent to do so. If both agents choose action $F$ simultaneously, then one of them will be randomly picked as the joint decision-maker for both agents.

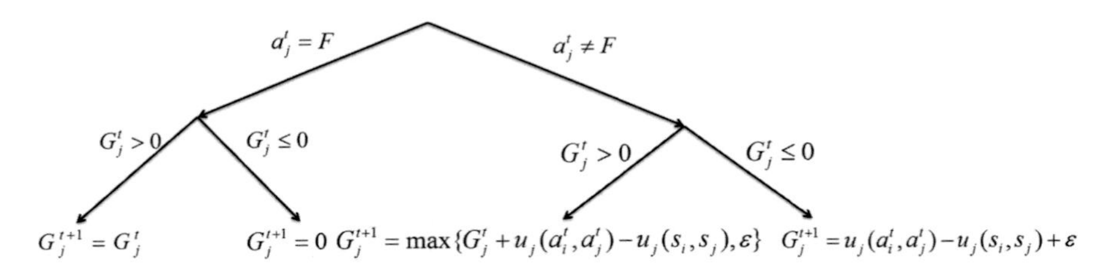


**Fig. 6.5** The rules of calculating the opponent $j$'s accumulated gain $G_j^t$ by each round $t$. Each path in the tree represents one case for calculating $G_j^t$

We can see that the opponent $j$ may obtain higher payoff by deviating from action $F$ to some action from $A_j$. Therefore a TaFSO agent $i$ needs to enable its opponent $j$ to learn to be aware that entrusting the TaFSO agent to make decisions is its best choice. To achieve this, the TaFSO agent $i$ will teach its opponent by punishment if the opponent $j$ chooses actions from $A_j$. To make it effective, the punishment must exceed the profit of deviating. In other words, the opponent $j$'s any possible gain from its deviation has to be wiped out. The TaFSO agent $i$ keeps the record of the opponent $j$'s accumulated gains $G_j^t$ from deviation by each round $t$ and updates $G_j^t$ as follows (Fig. 6.5):

- If the opponent $j$ entrusts the TaFSO agent $i$ to make decision for itself (i.e., $a_j^t = c$) and also its current gain $G_j^t \geq 0$, then its gain remains the same in the next round $t + 1$.
- If the opponent $j$ asks the TaFSO agent $i$ to make decision for itself (i.e., $a_j^t = c$) and also its current gain $G_j^t < 0$, this indicates it suffers from previous deviations. In this case, we forgive its previous deviations and set $G_j^{t+1} = 0$.
- If the opponent $j$ chooses its action independently and also $G_j^t > 0$, then its gain is updated as $G_j^{t+1} = \max\{G_j^t + u_j(a_i^t, a_j^t) - u_j(s_i, s_j), \epsilon\}$. If it obtains higher payoff than that in the optimal outcome $(s_i, s_j)$ from this deviation, then its gain will be increased, and vice versa. Besides, its total gain by round $t + 1$ cannot become smaller than zero since it deviates in the current round.
- If the opponent $j$ chooses its action independently and also $G_j^t \leq 0$, its gain is updated as $G_j^{t+1} = u_j(a_i^t, a_j^t) - u_j(s_i, s_j) + \epsilon$. That is, we forgive the opponent $j$ since it learns the lesson by itself by suffering from previous deviations ($G_j^t \leq 0$), and its previous gain is counted only as $\epsilon$.

Based on the above updating rules of $G_j^t$, the TaFSO agent needs to determine which action is chosen to punish the opponent $j$. To do this, the TaFSO agent $i$ keeps a teaching function $T_i^t(a)$ for each action $a \in A_i \cup \{F\}$ in each round $t$, indicating the action's punishment degree. From the Folk theorem [12], we know that the minimum payoff the TaFSO agent can guarantee that the opponent $j$ receives via punishment is its minimax payoff minimax$_j$. If $G_j^t >$ minimax$_j$, we can only expect to exert minimax$_j$ amount of punishment on the opponent $j$ since the opponent $j$ can always obtain at least minimax$_j$ by playing its maxmin strategy; if $G_j^t \leq$ minimax$_j$, then punishing the opponent $j$ by the amount of $G_j^t$ is already enough. Therefore the punishment degree of action $a$ in round $t$ is evaluated as the difference between the expected punishment on the opponent $j$ by choosing $a$ and the minimum value between $G_j^t$ and the punishment degree under the minimax strategy. Formally the teaching function $T_i^t(a)$ is defined as follows:

$$T_i^t(a) = u_j(s_1, s_2) - E[u_j(a, b)] - \min\{G_j^t, u_j(s_1, s_2) - \text{minimax}_j\} \tag{6.1}$$

where $\text{minimax}_j$ is the minimax payoff of the opponent $j$ and $E[u_j(a, b)]$ is the expected payoff that the opponent $j$ will obtain if the TaFSO agent $i$ chooses action $a$ based on the past history. Formally, $E[u_j(a, b)]$ can be expressed as follows:

$$E[u_j(a, b)] = \sum_{b \in A_j} (\text{freq}_j(b) \times u_j(a, b)) \tag{6.2}$$

where $\text{freq}_j(b)$ is the approximated probability that the opponent $j$ will play action $b$ next round based on the past history.

If $T_i^t(a) \geq 0$, it means it is sufficient to choose action $a$ to punish the opponent $j$. We can see that there can exist multiple candidate actions for punishing the opponent $j$. If $T_i^t(a) < 0$, $\forall a \in A_i$, then we only choose the action with the highest $T_i^t(a)$ as the candidate action. Overall the set $C_i^t$ of candidate actions for punishment is obtained based on TaFSO's *oracle* strategy in each round $t$. Based on this information, the TaFSO agent $i$ chooses an action from this set $C_i^t$ to punish its opponent according to its *follower* strategy, which will be introduced next.

#### 6.1.2.2 *Follower* Strategy in TaFSO

The *follower* strategy in TaFSO is used to determine the best response to the strategy of the opponent if the opponent chooses its action from its original action space. Here we adopt the $Q$-learning algorithm [5] as the basis of the *follower* strategy. Specifically the TaFSO agent $i$ holds a $Q$-value $Q_i^t(a)$ for each action $a \in A_i \cup \{F\}$ and gradually updates its $Q$-value $Q_i^t(a)$ for each action $a$ based on its own payoff and action in each round. The $Q$-value update rule for each action $a$ is as follows:

$$Q_i^{t+1}(a) = \begin{cases} Q_i^t(a) + \alpha_i(u_i^t(O) - Q_i^t(a)) & \text{if } a \text{ is chosen in round } t \\ Q_i^t(a) & \text{otherwise} \end{cases}, \tag{6.3}$$

where $u_i^t(O)$ is the payoff agent $i$ obtains in round $t$ under current outcome $O$ by taking action $a$. Besides, $\alpha_i$ is the learning rate of agent $i$, which determines how much weight we give to the newly acquired payoff $u_i^t(O)$, as opposed to the old $Q$-value $Q_i^t(a)$. If $\alpha_i = 0$, agent $i$ will learn nothing and the $Q$-value will be constant; if $\alpha_i = 1$, agent $i$ will only consider the newly acquired information $u_i^t(O)$.

In each round $t$, the TaFSO agent $i$ chooses its action based on the $\epsilon$-greedy exploration mechanism as follows. With probability $1-\epsilon$, it chooses the action with the highest $Q$-value from the set $C_i^t$ of candidate actions and chooses one action randomly with probability $\epsilon$ from the original action set $A_i \cup F$. The value of $\epsilon$ controls the exploration degree during learning. It initially starts at a high value and decreased gradually to nothing as time goes on. The reason is that initially the approximations of both the teaching function and the $Q$-value function are inaccurate and the agent has no idea of which action is optimal; thus the value of $\epsilon$ is set to a relatively high value to allow the agent to explore potential optimal actions. After enough explorations, the exploration has to be stopped so that the agent will focus on only exploiting the action that has shown to be optimal before.

#### 6.1.2.3 Overall Algorithm of TaFSO

The overall algorithm of TaFSO is sketched in Algorithm 7, and it combines the *oracle* and *follower* elements we previously described. The only difference is that a special rule (line 5–9) is added to identify whether the opponent is adopting TaFSO or not for the case of self-play. If the opponent also adopts TaFSO, it is equivalent to the reduced case that both agents alternatively decide the joint action and thus the pre-calculated optimal outcome $(s_1, s_2)$ is always achieved. Otherwise, during each round, the TaFSO agent first determines the optimal joint action and also the set of candidate actions based on its *oracle* strategy and then chooses an action to

**Algorithm 7** Overall algorithm of TaFSO

1: Initialize $G_j^t$, $Q(a)$, $\forall a \in A_i \cup F$
2: Observe the game $G$; calculate the optimal outcome $(s_1, s_2)$.
3: **for** each round $t$ **do**
4: Compute the set $C_j^t$ of candidate actions.
5: **if** t = 1 **then**
6: Choose action $F$.
7: **else**
8: **if** $a_j^{t-1} = F$ **then**
9: Choose action $F$.
10: **else**
11: Choose an action $a_i^t$ according to the *follower* strategy in Sect. 6.1.2.2.
12: **end if**
13: **end if**
14: **if** it becomes the joint decision-maker **then**
15: Choose the precomputed optimal outcome $(s_1, s_2)$ as the joint decision.
16: Update $G_j^t$ based on the update rule in Sect. 6.1.2.1.
17: **else**
18: Update $Q(a)$, $\forall a \in A_i \cup F$ following Eq. 6.3 after receiving the reward of the joint outcome $(a_i^t, a_j^t)$.
19: Update $G_j^t$ based on the update rule in Sect. 6.1.2.1.
20: **end if**
21: **end for**

execute following its *follower* strategy. The outcome of each round depends on the joint action of the TaFSO agent and its opponent, and also the $Q$-values and $G_j^t$ of the TaFSO agent will be updated accordingly (Line 14–20).

Next we make the following observations for the TaFSO algorithm from the *oracle* and *follower* strategies' perspectives.

#### Efficiency of the *Oracle* Strategy

In TaFSO, the teaching goal is to let the opponent be aware that entrusting the TaFSO agent to make decisions for itself is in its best interest. The opponent is always rewarded by the payoff in the optimal outcome $(s_1, s_2)$ when it chooses action $F$ and punished to wipe out its gain whenever it deviates by choosing action from its original action space. In this way, it prevents the occurrence of mistaken punishment and the opponent never wrongly perceives the punishment signal compared with previous work [8], thus making the teaching process more efficient. Besides, it is also easy to see that there is no mis-coordination when there exist multiple socially optimal outcomes, since the socially optimal outcome is always chosen by the TaFSO agent in a deterministic way (described in Sect. 6.1.2.1). Overall it is expected that a rational opponent can always be incentivized to choose action $F$ as long as its payoff under the socially optimal outcome assigned by the TaFSO agent is higher than its minimax payoff, which will be verified through extensive simulations in Sect. 6.2.3.

#### Efficiency of the *Follower* Strategy

From the *oracle* strategy, a set of candidate actions suitable for punishment is obtained based on the teaching function in Eq. 6.1. Different from trigger strategy, the teaching function predicts the opponent's next-round action based on the past history, instead of assuming that the opponent always takes the maxmin strategy. This is more reasonable and efficient since the opponent does not necessarily choose the maxmin strategy, and it is highly likely that there exist multiple action choices that are all sufficient to wipe out any possible gain of the opponent from past deviation.

According to the *follower* strategy, the TaFSO agent learns the relative performance of different actions (their $Q$-values) against the opponent. Given the set of candidate actions obtained from the teaching function, the TaFSO agent always chooses the action in its own best interest from the candidate actions through exploration and exploitation mechanism. In this way, the TaFSO agent can reduce its own punishment cost as much as possible when still guaranteeing that it is sufficient to exert punishment on its opponent. In contrast, an agent adopting trigger strategy always picks the minimax strategy to punish its opponent in a deterministic way without taking into consideration its punishment cost, which thus may make the teaching process quite inefficient.

### *6.1.3 Experimental Simulations*

In this section, we present the experimental results in two parts. First, in Sect. 6.1.3.1, we consider the case of playing against the class of best-response learners following the learning environment description in Sect. 6.1.1, which is the main focus of our current work. The learning performance of the TaFSO strategy against different best-response learners in terms of coordinating on socially optimal outcomes is evaluated by comparing with previous work, the SPaM strategy [8], under different testbeds. In the second part (Sect. 6.1.3.2), we focus on the case of self-play and compare the performance of TaFSO under self-play with previous work [15, 16] using the testbed adopted in [15] based on a number of commonly adopted evaluation criteria [15].

#### 6.1.3.1 Against Best-Response Learners

We evaluate the performance of TaFSO against the opponents adopting a variety of different best-response strategies.[3] Specifically here we consider the opponent may adopt one of the following best-response strategies: $Q$-learning [5], WoLF-PHC [6], and fictitious play (FP) [7]. We compare the performance of TaFSO with SPaM [8] against the same set of opponents. The first set of testbed we adopt here is the same as the one in [8] by using the following three representative games: prisoner's dilemma game (Fig. 6.1), game of chicken (Fig. 6.6), and tricky game (Fig. 6.7).

**Fig. 6.6** Payoff matrix for game of chicken

| 1's payoff, 2's payoff | | Player 2's action | |
|---|---|---|---|
| | | C | D |
| Player 1's action | C | 4, 4 | 2, 5 |
| | D | 5, 2 | 0, 0 |

**Fig. 6.7** Payoff matrix for tricky game

| 1's payoff, 2's payoff | | Player 2's action | |
|---|---|---|---|
| | | C | D |
| Player 1's action | C | 0, 3 | 3, 2 |
| | D | 1, 0 | 2, 1 |

[3]In our current learning context, we assume that the agents using best-response strategies will simply choose the joint action pair with the highest payoff for itself when it becomes the decision-maker for both agents.

**Table 6.1** The agents' average payoffs using TaFSO and SPaM strategies against a number of best-response learners in three representative games

| Average payoffs | Prisoner's dilemma game | Game of chicken | Tricky game |
|---|---|---|---|
| TaFSO vs. *Q*-learning | 3.0 | 4.0 | 2.5 |
| SPaM vs. *Q*-learning | 2.46 | 3.46 | 2.15 |
| TaFSO vs. WoLF-PHC | 3.0 | 4.0 | 2.5 |
| SPaM vs. WoLF-PHC | 2.45 | 3.48 | 2.1 |
| TaFSO vs. FP | 3.0 | 4.0 | 2.5 |
| SPaM vs. FP | 2.5 | 3.6 | 2.17 |

For the prisoner's dilemma game, the socially optimal outcome is $(C, C)$, in which both agents receive a payoff of 3. For the game of chicken, the target solution is also $(C, C)$, in which both agents obtain a payoff of 4. In the tricky game, the socially optimal solution is $(C, D)$, and the agents' average payoffs are 2.5. Table 6.1 shows the agents' average payoffs when both TaFSO and SPaM strategies are adopted to repeatedly play the above representative games against different opponents.[4] We can see that the agents adopting TaFSO can always receive the average payoffs corresponding to the socially optimal outcomes for different games. For comparison, for those cases when SPaM is adopted, the agents' average payoffs are relatively lower than the maximum value of the socially optimal outcomes. The main reason is that the opponent agents adopting the best-response strategies may wrongly perceive the punishment signals of the SPaM agent and thus result in mis-coordination occasionally.

Next we further evaluate the performance of TaFSO strategy under a larger class of games, the 57 conflicting-interest game matrices with strict ordinal payoffs, as the testbed. This testbed was proposed by Brams in [17], which has been widely adopted to evaluate the learning performance of different learning algorithms [15, 18]. Generally conflicting-interest games are those games in which the players disagree on their most-preferred outcomes. These 57 game matrices cover all the structurally distinct two-player two-action conflicting-interest games, and we simply use the rank of each outcome as its payoff for each agent. All the 57 games are listed in Appendix A. For the nonconflicting interest (common interest) games, it is trivial since there always exists a Nash equilibrium that both players prefer most and also is socially optimal.

For these 57 conflicting-interest games, Table 6.2 shows whether there exists a SOSNE for each game and also whether a socially optimal outcome can be converged to when a TaFSO agent as the row player interacts against different types of best-response learners as the column player. First, from Table 6.2 we can see that the agents can always learn to converge to the SOSNE outcome for those games

[4]Note that only the payoffs obtained after 500 rounds are counted here since at the beginning the agents may achieve very low payoffs due to initial explorations. The results are averaged over 50 runs.

**Table 6.2** Results under the 57 conflicting-interest games when the TaFSO agent is the row player and its opponent is the column player

| Converge to socially optimal outcome | Games 1–9 | Game 10 | Games 11–25 | Game 26 | Games 27–41 | Game 42 | Games 43–44 | Game 45 | Games 46–57 |
|---|---|---|---|---|---|---|---|---|---|
| Existence of SOSNE | Yes | No | Yes | No | Yes | No | Yes | No | Yes |
| TaFSO vs. $Q$-learning | Yes | No | Yes | Yes | Yes | Yes | Yes | No | Yes |
| TaFSO vs. WoLF-PHC | Yes | No | Yes | Yes | Yes | Yes | Yes | No | Yes |
| TaFSO vs. FP | Yes | No | Yes | Yes | Yes | Yes | Yes | No | Yes |

**Table 6.3** Results under the 57 conflicting-interest games when the SPaM agent is the row player and the rational opponent is the column player

| Converge to socially optimal outcome | Games 1–9 | Game 10 | Games 11–31 | Game 32 | Games 23–34 | Game 35 | Games 36–44 | Game 45 | Games 46–57 |
|---|---|---|---|---|---|---|---|---|---|
| SPaM vs. $Q$-learning | Yes | No | Yes | No | Yes | No | Yes | No | Yes |
| SPaM vs. WoLF-PHC | Yes | No | Yes | No | Yes | No | Yes | No | Yes |
| SPaM vs. FP | Yes | No | Yes | No | Yes | No | Yes | No | Yes |

with a SOSNE outcome. However, it is also interesting to notice that for half of games without a SOSNE outcome, the agents can also successfully learn to converge to a socially optimal outcome instead of inefficient Nash equilibrium. There are only two games (game 10 and game 45) in which the agents fail to converge to a socially optimal outcome. Let us look closely into these two games shown in the Appendix A. It is easy to check that the minimax payoff profiles of these two games are $(2, 3)$ and $(3, 3)$, respectively; however, the socially optimal outcome of both games is $(4, 2)$. The rational opponent (as column player)'s payoff under the socially optimal outcome is lower than its minimax payoff; thus the punishment and reward mechanism of a TaFSO agent fails when it plays against a rational opponent under these two games. The punishment and reward mechanism of the TaFSO strategy only works when the opponent's minimax payoff is smaller than its payoff under the socially optimal outcome.

Table 6.3 shows the results of converging to socially optimal outcomes for the 57 structurally distinct games when the SPaM agent (row player) plays against different rational opponents (column player). We can see that the agents fail to reach socially optimal outcome under four games: game 10, 32, 35, and 45. For the games 10 and 45, the failure reason is similar to the case of TaFSO agent: the minimax payoff of the opponent as the column player is larger than that under the socially optimal outcome; thus a rational opponent has no incentive to coordinate on the socially optimal outcome. However, the agents also fail under another two games, games 32 and 35, which is caused by the rational opponent's misperception of the punishment signal from the SPaM agent. Taking game 32 for example, the SPaM agent expects to coordinate on the outcome $(D, D)$, while the rational opponent may deviate by choosing action $C$ to increase its payoff. In this case, the SPaM agent would choose action $C$ to punish the opponent, which would reinforce the opponent to choose action $C$, since choosing action $D$ is worse for itself when the SPaM agent chooses action $C$ (performing punishment). In contrast, there is no misperception of the punishment signal under the action entrustment mechanism, and the TaFSO agent can successfully coordinate on the socially optimal outcomes under both games 32 and 35 when playing against all rational opponents considered.

For asymmetric games, since the agents adopt different strategies, the learning results can be quite different if their roles (row player or column player) change. To this end, we first reverse the roles of the TaFSO (or SPaM) agent and the rational opponent and reevaluate the performance of both TaFSO and SPaM agents playing against different rational opponents. Table 6.4 shows the convergence results under the same 57 conflicting-interest games when different rational opponents (as the row player) play against the TaFSO agent (as the column player). Similar to the results in Table 6.2, the agents can always converge to the SOSNE outcome whenever the game being played has such an outcome. However, for those games without a SOSNE outcome, the agents fail to converge to a socially optimal outcome in game 26 and game 42. This can be explained in a similar way as the previous analysis for the results in Table 6.2. The only difference is that the roles of the TaFSO agent and its best-response learning opponents are reversed. Take game 26 as an example, the minimax payoff profile is (3, 2) and the payoff profile corresponding to the socially optimal outcome is (2, 4). Thus it is easy to see that the best-response learners have no incentive to learn to choose action $F$ since its payoff under this socially optimal outcome (i.e., 2) is lower than its minimax payoff (i.e., 3).

Table 6.5 shows the results under the 57 games when different rational opponent (row player) play against the SPaM agent (column player). First, we can observe that they fail to achieve socially optimal outcome under both game 26 and game 42, similar to the case of the TaFSO agent. This can be explained by the same reason as

**Table 6.4** Results under the 57 conflicting-interest games when the TaFSO agent acts as the column player and its opponent as the row player

| Converge to socially optimal outcome | Games 1–9 | Game 10 | Games 11–25 | Game 26 | Games 27–41 | Game 42 | Games 43–44 | Game 45 | Games 46–57 |
|---|---|---|---|---|---|---|---|---|---|
| Existence of SOSNE | Yes | No | Yes | No | Yes | No | Yes | No | Yes |
| $Q$-learning vs. TaFSO | Yes | Yes | Yes | No | Yes | No | Yes | Yes | Yes |
| WoLF-PHC vs. TaFSO | Yes | Yes | Yes | No | Yes | No | Yes | Yes | Yes |
| FP vs. TaFSO | Yes | Yes | Yes | No | Yes | No | Yes | Yes | Yes |

**Table 6.5** Results under the 57 conflicting-interest games when the SPaM agent is the column player and the rational opponent is the row player

| Converge to socially optimal outcome | Games 1–25 | Games 26–28 | Games 29–31 | Game 32 | Games 33–41 | Game 42 | Games 43–47 | Games 48–49 | Games 50–57 |
|---|---|---|---|---|---|---|---|---|---|
| $Q$-learning vs. SPaM | Yes | No | Yes | No | Yes | No | Yes | No | Yes |
| WoLF-PHC vs. SPaM | Yes | No | Yes | No | Yes | No | Yes | No | Yes |
| FP vs. SPaM | Yes | No | Yes | No | Yes | No | Yes | No | Yes |

before: the rational agent as the row player can achieve higher payoff under minimax outcome than that under the socially optimal outcome; thus it has no incentive to cooperate. Besides, there also exist other three games (games 32, 48, and 49) under which the agents fail to achieve socially optimal outcomes. We can notice that all these three games share the common characteristic that a rational opponent as the row player has the incentive to deviate from the socially optimal outcome, and also the punishment signal from the SPaM agent will reinforce the rational opponent to stay there. Accordingly, the agents will eventually converge to an outcome which is not socially optimal (not necessary to be a Nash equilibrium). In contrast, there is no such misperception of punishment signal from the TaFSO agent due to the introduction of the action entrustment mechanism, and thus the agents can always successfully converge to socially optimal outcomes under all these three games.

Overall, based on the above results in Tables 6.2, 6.3, 6.4, and 6.5, we can see that under the action entrustment mechanism, the TaFSO agent can induce rational opponents to converge to socially optimal outcomes for more games compared with the SPaM agent.

#### 6.1.3.2 Under Self-Play (*Oracle* Against *Oracle*)

In this section we compare the performance of TaFSO with CJAL [15], action revelation [16], and WOLF-PHC [6] in two-player games under self-play. Both players play each game repeatedly for 2000 time steps with learning rate of 0.6. The exploration rate starts at 0.3 and gradually decreases by 0.0002 each time step. For all previous strategies, the same parameter settings as those in their original papers are adopted.

Here we again use the 57 conflicting-interest game matrices with strict ordinal payoffs proposed by Brams in [17] as the testbed for evaluation. For the nonconflicting interest games, it is trivial since there always exists a Nash equilibrium in the single-stage game that both players prefer most and also is socially optimal. It is easy for the agents to learn to converge to this socially optimal Nash equilibrium for all the learning strategies we consider here, and thus this type of games is not considered.

The performance of each approach is evaluated in self-play on these 57 conflicting-interest games, and we compare their performance based on the following three criteria [15]. The comparison results are obtained by averaging over 50 runs across all the 57 conflicting-interest games.

**Utilitarian Social Welfare** The utilitarian collective utility function $sw_U(P)$ for calculating utilitarian social welfare is defined as $sw_U(P) = \sum_i^n p_i$, where $P = \{p_i\}_i^n$ and $p_i$ is the actual payoff agent $i$ obtains when the outcome is converged. Since the primary learning goal is to achieve socially optimal outcomes, utilitarian social welfare is the most desirable criterion for performance evaluation.

**Table 6.6** Performance comparison with CJAL, action revelation, and WOLF-PHC using the testbed in [17]

| | Utilitarian social welfare | Nash social welfare | Success rate |
|---|---|---|---|
| TaFSO (our strategy) | 6.45 | 10.08 | 0.96 |
| CJAL [15] | 6.14 | 9.25 | 0.86 |
| Action revelation [16] | 6.17 | 9.30 | 0.81 |
| WOLF-PHC [6] | 6.03 | 9.01 | 0.75 |
| Nash | 6.05 | 9.04 | 0.75 |

**Nash Social Welfare** Nash social welfare is also an important evaluation metrics in that it strikes a balance between maximizing utilitarian social welfare and achieving fairness. Its corresponding utility function $sw_N(P)$ is defined as $sw_N(P) = \prod_i^n p_i$, where $P = \{p_i\}_i^n$ and $p_i$ is the actual payoff agent $i$ obtains when the outcome is converged. On one hand, Nash social welfare reflects utilitarian social welfare. If any individual agent's payoff decreases, the Nash social welfare also decreases. On the other hand, it also reflects the fairness degree between individual agents. If the total payoff is a constant, then Nash social welfare is maximized only if the payoffs is shared equally among agents.

**Success Rate** Success rate is defined as the percentage of times that the socially optimal outcome (maximizing utilitarian social welfare) is converged at last.

The comparison results based on these three criteria are shown in Table 6.6. We can see that TaFSO outperforms all these three strategies in terms of the above criteria. Players using ToFSO can obtain utilitarian social welfare of 6.45 and Nash social welfare of 10.08 with success rate of 0.96, which are higher than all the other three approaches. We also provide the average Nash equilibrium payoffs and its success rate in terms of achieving socially optimal outcomes for all the 57 games. Note that the performance of WOLF-PHC approach is the worst since this approach is designed for achieving Nash equilibrium in single-stage game only which often does not coincide with socially optimal solution (only 75 % of the games satisfy this requirement). For action revelation and CJAL, they both fail in certain types of games, e.g., the prisoner's dilemma game. For action revelation, self-interested agent can always exploit the action revelation mechanism and have the incentive to choose defection $D$, thus leading the outcome to converge to mutual defection; for CJAL, it requires the agents to randomly explore for a finite number of rounds $N$ first, and the probability of converging to mutual cooperation tends to 1 only if the value of $N$ approaches infinity. Besides, it only works when the payoff structure of the prisoner's dilemma game satisfies certain condition [15]. For example, consider the two different versions of the prisoner's dilemma game in Figs. 6.8 and 6.9. For both WOLF-PHC and action revelation, the agents always converge to the pure strategy Nash equilibrium $(D, D)$; for CJAL, the agents can successfully learn to converge to the socially optimal outcome $(C, C)$ for the first prisoner's dilemma game, while fail to converge to $(C, C)$ for the second one [15]. In contrast, the agents

**Fig. 6.8** Payoff matrix for prisoner's dilemma game: version 1

| 1's payoff, 2's payoff | | Player 2's action | |
|---|---|---|---|
| | | C | D |
| Player 1's action | C | 3, 3 | 0, 5 |
| | D | 5, 0 | 1, 1 |

**Fig. 6.9** Payoff matrix for prisoner's dilemma game: version 2

| 1's payoff, 2's payoff | | Player 2's action | |
|---|---|---|---|
| | | C | D |
| Player 1's action | C | 3, 3 | 0, 5 |
| | D | 5, 0 | 2, 2 |

using TaFSO can always coordinate on the socially optimal outcome $(C, C)$ for both instances of the prisoner's dilemma game under self-play.

## 6.2 Achieving Socially Optimal Solutions in the Social Learning Framework

Previous section considers the problem of achieving socially optimal outcomes under the framework of the two-agent repeated interactions; however, in practice, there may exist a large number of agents interacting with each other. Therefore, next we consider the same problem of achieving socially optimal outcomes under the second learning framework—*social learning* framework, which usually involves a large number of agents interacting with one another [3].

One approach of addressing this problem is to enforce the agents to behave in a socially rational way—aiming at maximizing the sum of all agents' utilities when making their own decisions. However, as mentioned in previous work [19], this line of approach suffers from a lot of drawbacks. First it may greatly increase the computational burdens of individual agents, since each agent needs to consider all agents' interests into consideration when it makes decisions. Besides, it may become infeasible to enforce the agents to act in a socially rational manner if the system is open in which we have no control on the behaviors of all agents in the system. To solve these problems, one natural direction is considering how we can incentivize the individually rational agents to act toward coordinating on socially optimal solutions. A number of works [19–22] have been done in this direction by designing different interaction mechanisms of the system while the individual rationality of the interacting agents is maintained and respected at the same time. One common drawback of previous work is that certain amount of global information is required to be accessible to each individual agent in the system, and also there still exists certain percentage of agents that are not able to learn to coordinate on socially optimal outcomes (SOs).

To this end, we propose inserting a number of influencer agents into the system to incentivize the rest of individually rational agents to behave in a socially rational way. The concept of influencer agent is first proposed by Franks et al. [23] and has been shown to be effective in promoting the emergence of high-quality norms in the linguistic coordination domain [24]. In general, an influencer agent is an agent with desirable convention or prosocial behavior, which is usually inserted into the system by the system designer aiming at manipulating those individually rational agents into adopting desirable conventions or behaviors. To enable that the influencer agents exert effective influences on individually rational agents, we consider an interesting variation of sequential play by allowing entrusting decision to others similar to previous work [1]. During each interaction between each pair of agents, apart from choosing an action from its original action set, each agent is also given the option of choosing to entrust its interacting partner to make a joint decision for both agents. It should be noted that such a decision to entrust the opponent is completely voluntary; hence the autonomy and rationality of an agent are well respected and maintained. The influencer agents are socially rational in the sense that they will always select an action pair that corresponds to a socially optimal outcome should it become the joint decision-maker. Besides, each agent is allowed to choose to interact with an influencer agent or an individually rational agent, and then it will interact with a randomly chosen agent of that type. Each agent (both individually rational and influencer agents) uses a rational learning algorithm to make its decisions in terms of which type of agent to interact with and which action to choose and improves its learning policy based on the reward it receives from the interaction. We evaluate the performance of the learning framework in two representative types of games: PD game and anti-coordination (AC) game. Simulation results show that a small proportion of influencer agents can efficiently incentivize most of the purely rational agents to coordinate on the socially optimal outcomes and better performance in terms of the average percentage of socially optimal outcome attained can be achieved compared with that of previous work [22].

### *6.2.1 Social Learning Environment and Goal*

The general question we are interested in is how individually rational agents can learn to coordinate with one another on desirable outcomes through repeated pairwise interactions. In particular, we aim to achieve socially optimal outcomes, under which the utilitarian social welfare (i.e., the sum of all agents' payoffs) is maximized. At the same time, we desire that the rationality and autonomy of individual agents be maintained. In other words, the agents should act independently in a completely individually rational manner when they make decisions. This property is highly desirable particularly when the system is within an open, unpredictable environment, since the system implemented with this kind of property can largely withstand the exploitations of selfish agents designed by other parties.

Specifically, we study the learning problem toward socially optimal solutions in the context of a population of agents as follows. In each round each agent chooses to interact with another agent (i.e., to play a game with that agent), which is constrained by the interaction protocol of the system. Each agent learns concurrently over repeated interactions with other agents in the system. The interaction between each pair of agents is formulated as a two-player normal-form game, which will be introduced later. We assume that the agents are located in a distributed environment and there is no central controller for determining the agents' behaviors. Each agent can only know its own payoff during each interaction and makes decisions autonomously.

Following previous work [19, 21, 22], we focus on two-player two-action symmetric games for modeling the agents' interactions, which can be classified into two different types. For the first type of games, the agents need to coordinate on the outcomes with identical actions to achieve socially rational outcomes. One representative game is the well-known PD game (see Fig. 6.10), in which the socially optimal outcome is $(C, C)$; however, choosing action $D$ is always the best strategy for any individually rational agent. The second type of games requires the agents to coordinate on outcomes with complementary actions to achieve socially optimal outcomes. Its representative game is AC game (see Fig. 6.11), in which either outcomes $(C, D)$ or $(D, C)$ is socially optimal. However, the row and column agents prefer different outcomes and thus it is highly likely for individually rational agents to fail to coordinate (achieving inefficient outcomes $(C, C)$ or $(D, D)$). For both types of games, we are interested in investigating how the individually rational agents can be incentivized to learn to efficiently coordinate on the corresponding socially optimal outcomes.

**Fig. 6.10** Prisoner's dilemma (PD) game satisfying the constraints of $T > R > P > S$ and $2R > T + S > 2P$

| A's payoff, B's payoff | | Agent B's action | |
|---|---|---|---|
| | | C | D |
| Agent A's action | C | R,R | S,T |
| | D | T,S | P,P |

**Fig. 6.11** Anti-coordination (AC) game satisfying the constraints of $H > L$

| A's payoff, B's payoff | | Agent B's action | |
|---|---|---|---|
| | | C | D |
| Agent A's action | C | L,L | H,H |
| | D | H,H | L,L |

## 6.2.2 *Learning Framework*

We first give a background introduction on the concept of influencer agent and how it can be applied for solving our problem in Sect. 6.2.2.1. Then we describe the interaction protocol within the framework in Sect. 6.2.2.2. Finally the learning strategy the agents adopt to make decisions is introduced in Sect. 6.2.2.3.

### 6.2.2.1 Influencer Agent

The concept of influencer agent is firstly termed by Franks et al. [23], and there are also a number of previous works with similar ideas. In general, an influencer agent (IA) is an agent inserted into the system usually by the system designer in order to achieve certain desirable goals, e.g., emergence of efficient convention or norms [23]. Sen and Airiau [25] investigate and show that a small portion of agents with fixed convention can significantly influence the behavior of large group of selfish agents in the system in terms of which convention will be adopted in the system. Similarly Franks et al. [23] investigate the problem of how a small set of influencer agents adopting prefixed desirable convention can influence the rest of individually rational agents toward adopting the convention the system designer desires in the linguistic coordination domain [24].

Since we are interested in incentivizing individually rational agents to behave in the socially rational way, here we consider inserting a small number of influencer agents, which are socially rational, into the system. To enable the influencer agents to exert effective influence on individually rational agents' behaviors, we consider an interesting variation of sequential play by allowing entrusting decision to others similar to previous work [1]. During each interaction between each pair of agents, apart from choosing an action from its original action set, each agent is also given an additional option of asking its interacting partner to make the decision for both agents (denoted as choosing action $F$). If an agent $A$ chooses action $F$, while its interacting partner $B$ does not, agent $B$ will act as the leader to make the joint decision for them. If both agents choose action $F$ simultaneously, then one of them will be randomly chosen as the joint decision-maker. The influencer agents are socially rational in that they will always select the socially optimal outcome as the joint action pair to execute whenever it becomes the joint decision-maker. If there exist multiple socially optimal outcomes, then these socially optimal outcomes will be selected with equal probability. For those individually rational agents, we simply assume that they will always choose the outcome under which their own payoffs are maximized as the joint action for execution, whenever they are entrusted to make joint decisions.

#### 6.2.2.2 Interaction Protocol

From previous description, we know that there exist two different types of agents in the system: influencer agents (IA) and individually rational (or "selfish") agents (SA). In each round, each agent is allowed to choose which type of agent to interact with, and then it will interact with an agent randomly chosen from the corresponding set of agents. This is similar to the commonly used interaction model that the agents are situated in a fully connected network in which each agent randomly interacts with another agent each round [25, 26]. The only difference is that in our model the population of agents are divided into two groups and each agent is given the freedom to decide which group to interact with first, but the specific agent to interact with within each group is still chosen randomly. Our interaction model can better reflect the realistic scenarios in human society, since human can be classified into different groups according to their personality traits and different persons may have different preferences regarding which group of people they are willing to interact with.

Each agent uses a rational learning algorithm to make its decisions in terms of which type of agent to interact with and which action to choose and improves its policy based on the rewards it receives during the interactions. Besides, each agent chosen as the interacting partner also needs to choose an action to respond accordingly depending on the type of its interacting agent. We assume that during each interaction each agent only knows its own payoff and cannot have access to its interacting partner's payoff and action. The overall interaction protocol is shown in Algorithm 8.

#### 6.2.2.3 Learning Algorithm

For the individually rational agents, it is natural that they always choose the strategy which is a best response to its partner's current strategy in order to maximize its own payoff. If a learning algorithm has the property that it can converge to a policy that is a best response to the other players' policies when the other players' policies converge to stationary ones, then it is regarded as being rational [27]. A number of rational learning algorithms exist in the multiagent learning literature and here we adopt the $Q$-learning algorithm [5], which is the most commonly used. Specifically, each individually rational agent maintains two different sets

**Algorithm 8** Interaction protocol

```
1: for a fixed number of rounds do
2:     for each agent i in the system do
3:         determine which type of agents to interact with
4:         interact with one agent randomly chosen from the corresponding set of agents
5:         update its policy based on the reward received from the interaction
6:     end for
7: end for
```

of $Q$-tables: one corresponding to the estimates of the payoffs for actions for interacting with influencer agents, $Q_{IA}$, and the other corresponding to the estimates of the payoffs for the set of actions for interacting with individually rational agents, $Q_{SA}$. In the following discussion, $a_{IA}$ refers to an action when interacting with an influencer agent and $a_{SA}$ refers to an action when interacting with an individually rational agent. In each round $t$, an individually rational agent $i$ makes its decision (which type of agent to interact and which specific action to choose) based on the Boltzmann exploration mechanism as follows. Formally any action $a_{IA}$ belonging to the set of actions available for interacting with influencer agents is selected with probability

$$\frac{e^{Q_{IA}(a_{IA})/T}}{\sum_{a_{IA}} e^{Q_{IA}(a_{IA})/T} + \sum_{a_{SA}} e^{Q_{SA}(a_{SA})/T}}. \tag{6.4}$$

Any action $a_{SA}$ belonging to the set of actions available for interacting with individually rational agents is selected with probability

$$\frac{e^{Q_{SA}(a_{SA})/T}}{\sum_{a_{IA}} e^{Q_{IA}(a_{IA})/T} + \sum_{a_{SA}} e^{Q_{SA}(a_{SA})/T}}. \tag{6.5}$$

The temperature parameter $T$ controls the exploration degree during learning, and initially it is given a high value and decreased over time. The reason is that initially the approximations of the $Q$-value functions are inaccurate and the agents have no idea of which action is optimal; thus the value of $T$ is set to a relatively high value to allow the agents to explore potential optimal actions. After enough explorations, the exploration has to be stopped so that the agents can focus on exploiting the actions that has shown to be optimal before.

An individually rational agent selected as the interacting partner makes decisions depending on which type of agent it will interact with. If it interacts with another individually rational agent, then it will choose an action from the set of actions available for interacting with individually rational agents, and any action $a_{SA}$ is chosen with probability

$$\frac{e^{Q_{SA}(a_{SA})/T}}{\sum_{a_{IA}} e^{Q_{SA}(a_{SA})/T}}. \tag{6.6}$$

If it is chosen by an influencer agent, then it will pick an action from the set of actions available for interacting with influencer agents, and any action $a_{IA}$ is selected with probability

$$\frac{e^{Q_{IA}(a_{IA})/T}}{\sum_{a_{IA}} e^{Q_{IA}(a_{IA})/T}}. \tag{6.7}$$

After the interaction in each round $t$, each agent updates its corresponding $Q$-table depending on which type of agent it has interacted with. There are two different learning modalities available for performing update [26]: (1) multi-learning approach (for each pair of interacting agents, both agents update their $Q$-tables based on the payoffs they receive during interaction) and (2) mono-learning approach (only the agent who initiates the interaction updates its $Q$-table and its interacting partner does not update). In mono-learning approach, each agent updates its policy in the same speed, while in the multi-learning approach, some agents may learn much faster than others due to the bias of partner selection. We investigate both updating approaches and the effects of both updating approaches on the system-level performance will be shown in Sect. 6.2.3. Formally, each agent updates its $Q$-tables during each interaction as follows depending on which type of agent it has interacted with:

$$Q^{t+1}_{IA/SA}(a) = \begin{cases} Q^t_{IA/SA}(a) + \alpha_i(r^t_i - Q^t_{IA/SA}(a)) & \text{if } a \text{ is chosen in round } t \\ Q^t_{IA/SA}(a) & \text{otherwise} \end{cases}, \tag{6.8}$$

where $r^t_i$ is the reward agent $i$ obtains from the interaction in round $t$ by taking action $a$. $\alpha_i$ is the learning rate of agent $i$, which determines how much weight we give to the newly acquired reward $r^t_i$, as opposed to the old $Q$-value. If $\alpha_i = 0$, agent $i$ will learn nothing and the $Q$-value will be constant; if $\alpha_i = 1$, agent $i$ will only consider the newly acquired information $r^t_i$.

For influencer agents, we do not elevate their learning abilities above the rest of the population. They make decisions in the same way as individually rational agents. The only difference is that the influencer agents behave in a socially rational way in that they will always select the socially optimal outcome(s) as the joint action pair(s) to execute whenever it is selected as the joint decision-maker as described in Sect. 6.2.2.1.

### 6.2.3 Experimental Simulations

In this section, we present the simulation results showing how the influencer agents can significantly influence the population's behaviors toward social optimality in two types of representative games: PD game and AC game. All the simulations are performed in a population of 1000 agents. Following previous work [23], we consider 5 % of influencer agents in a population (i.e., 50 agents out of 1000) to be an appropriate upper bound of how many agents can be inserted into a system in practical application domains. However, for evaluation purpose, we perform simulations with the percentage of influencer agents up to 50 % in order to have a better understanding of the effects of influencer agents on the dynamics of the system. We first give an analysis of the effects of the number of influencer agents and different update modalities on the system-level performance in each game in

Sects. 6.2.3.1 and 6.2.3.2, respectively, and then compare the performance of our learning framework using influencer agents with that of previous work [22] in Sect. 6.2.3.3.

#### 6.2.3.1 Prisoner's Dilemma Game

For the PD game, we use the following setting: $R = 3, S = 0, T = 5, P = 1$. Figure 6.12 shows the average percentage of socially optimal outcome in the system when the number of influencer agents (IAs) inserted varies using mono-learning update. When no IAs are inserted into the system (i.e., a population of individually rational agents (SAs)), the percentage of socially optimal outcomes (SOs) achieved quickly reaches zero. This is obvious since choosing action $D$ is always the best choice for each individually rational agent when it interacts with another individually rational entity in PD game. By inserting a small amount of IAs with a proportion of 0.002 (2 IAs in the population of 1000 agents), we can see a significant gain in terms of the percentage of SOs attained up to 0.8. The underlying reason is that most of SAs are incentivized to voluntarily choose to interact with IAs and also select action $F$. Further increasing the number of IAs (to 10 IA agents) can significantly improve the speed of increase of percentage of SOs and also bring in small improvement of the percentage of SOs finally attained. We hypothesize that it is because some IAs' behaviors against SAs are not optimal when the number of

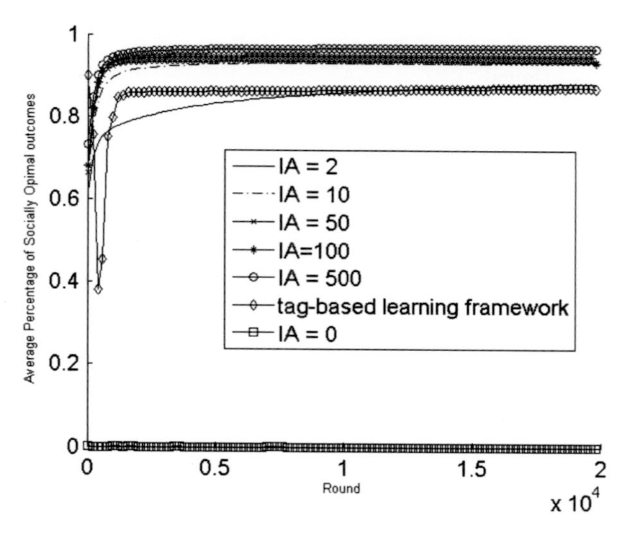


**Fig. 6.12** Average percentage of SOs with different number of IAs under mono-learning update (PD game)

IAs is small, and more IAs can successfully learn the optimal action against SAs when the number of IAs becomes larger. However, as the number of IAs is further increased, the increase in the final value of the proportion of SOs attained becomes less obvious. The reason is that with the number of IAs increasing, there is little additional benefit on the behaviors of IAs and the small amount of increase in the percentage of SOs is purely the result of the increase of the percentage of IAs itself.

Figure 6.13 shows the differences between updating using multi-learning and mono-learning approach on the system-level performance, i.e., the average percentage of SOs attained in the system, in PD game. From previous analysis, we have known that most SAs learn to interact with IAs and choose action $F$; thus IAs are given much more experience and opportunities to improve their policies against SAs under multi-learning approach. Accordingly it is expected that higher level of SOs can be achieved compared with that under mono-learning approach. It is easy to verify that the simulation results in Fig. 6.13 agree to our predications.

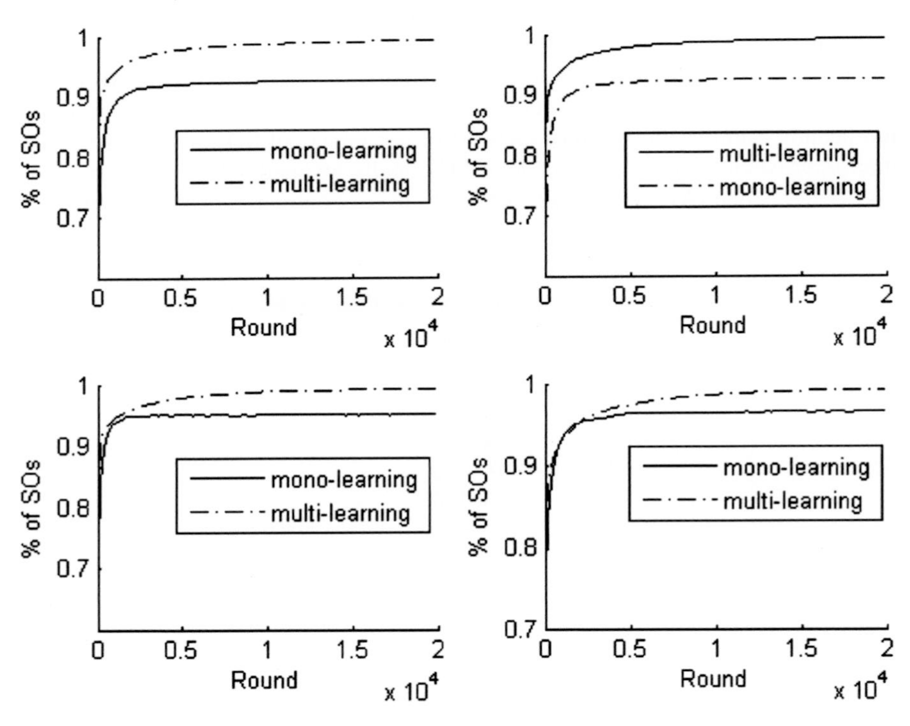


**Fig. 6.13** Mono-learning approach vs. multi-learning approach in PD game ($IA = 2, 20, 100, 500$)

#### 6.2.3.2 Anti-Coordination Game

In AC games, the following setting is used to evaluate the learning framework: $L = 1, H = 2$. Figure 6.14 shows the average proportion of socially optimal outcomes attained in the system when the number of influencer agents varies. Different from the PD game, when there is no influencer agents inserted in the system, most of the selfish agents (up to 85 %) can still learn to coordinate with each other on socially optimal outcomes. Initially the percentage of socially optimal outcomes is very high because most of the agents learn that choosing action $F$ is the best choice which prevents the occurrence of mis-coordination. However, gradually the agents realize that they can benefit more by exploiting those peer agents choosing action $F$ by choosing action $C$ or $D$ and thus this inevitably results in mis-coordination (i.e., achieving outcome $(C, C)$ or $(D, D)$) when these exploiting agents interact with each other. Thus the average percentage of socially optimal outcomes gradually drops to around 85 %. Besides, the mis-coordination rate converges to around 15 %, which can be understood as the dynamic equilibrium that the system of agents has converged to, i.e., the percentages of agents choosing actions $C$, $D$, and $F$ are stabilized.

Significant increase in the average percentage of socially optimal outcome attained (up to almost 100 %) can be observed when a small amount of influencer agents (2 IAs out of 1000 agents) is inserted into the system. This can be explained by the fact that the SAs learn to entrust those IAs to make the joint decisions and also the IAs choose between the two socially optimal outcomes randomly. There

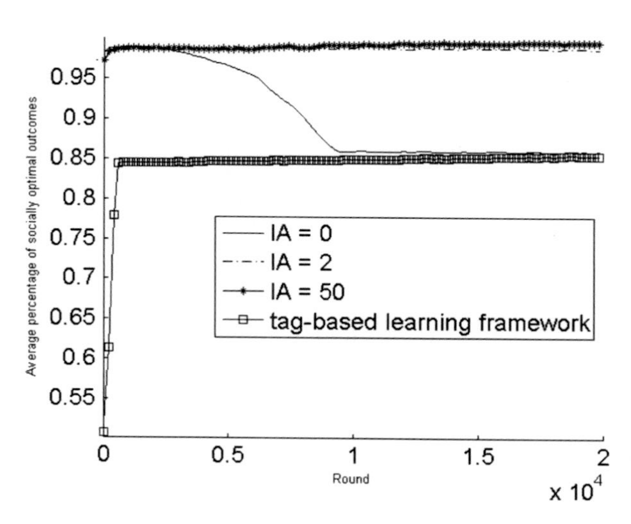


**Fig. 6.14** Average percentage of SOs with different number of IAs under mono-learning update (AC game)

is little incentive for the SAs to deviate since most of SAs have learned to interact with IAs and respond to SAs with action $C$ or $D$; thus there are no benefits to exploit other SAs due to high probability of mis-coordination. When the number of IAs is further increased (the number of IAs is 50), there is little performance increase in terms of the percentage of socially optimal outcomes achieved, and we only plot the case of $IA = 50$ for the purpose of clarity.

Figure 6.15 shows the differences between updating using multi-learning and mono-learning approach on the system-level performance, i.e., the average percentage of SOs attained in the system, in AC game with different number of IAs. Different from PD game, we can observe that slightly lower percentage of SOs is attained under multi-learning approach. We hypothesize that it is due to the fact that there exist two different socially optimal outcomes in AC game and thus it becomes easier for the agents to switch their policies between choosing these two outcomes and increase the chances of mis-coordination when they update their policies more frequently under multi-learning approach.

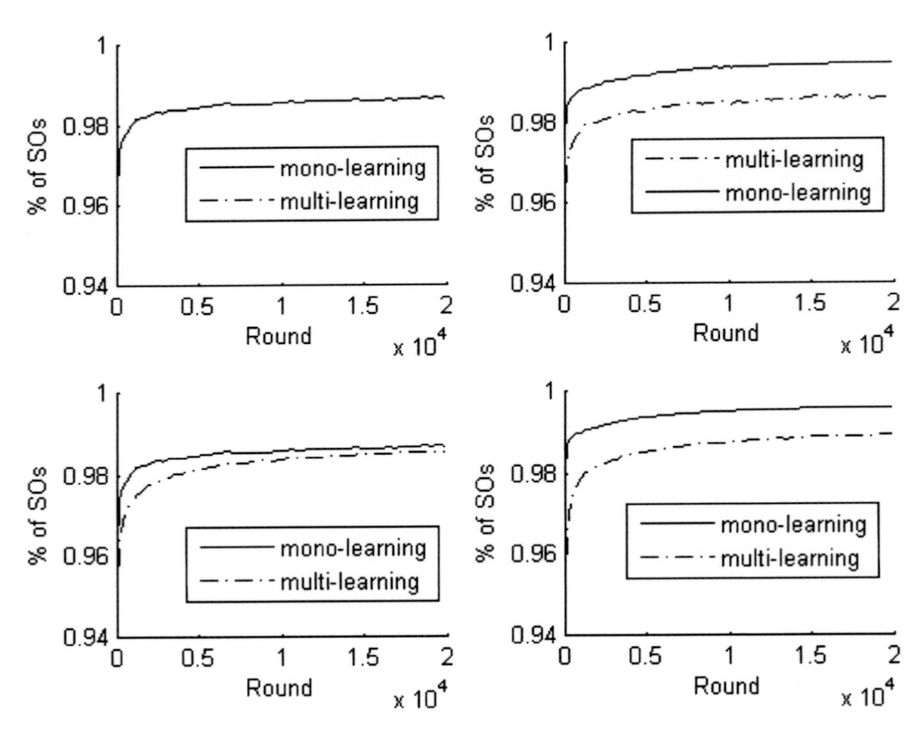


**Fig. 6.15** Mono-learning approach vs. multi-learning approach in AC game ($IA = 2, 10, 100, 500$)

#### 6.2.3.3 Comparison with Previous Work

We compare the performance of our learning framework using influencer agents with that of the tag-based learning framework [22] in both PD game and AC game. The number of influencer agents are set to 50 (5 % of the total number of agents) in our learning framework. The experimental setting for the tag-based learning framework follows the setting given in [22].

Figures 6.12 and 6.14 show the performance comparisons with the tag-based learning framework in PD game and AC game, respectively. For both cases, we can observe that there is a significant increase in the average percentage of SOs under our learning framework using influencer agents. Besides, the rate in which the agents converge to SOs is higher than that using the tag-based learning framework. The underlying reason is that under the tag-based learning framework, the agents learn their policies in a periodical way, and the coordination toward socially optimal outcomes requires at least two consecutive periods' adaptive learning between individual learning and social learning. Also our learning framework using IAs can better prevent the exploitations from SAs since the IAs act in the same way as SAs if their interacting partners do not choose action $F$. Accordingly, higher percentage of SAs can be incentivized to cooperate with IAs and thus higher percentage of socially optimal outcomes can be achieved.

## References

1. Hao JY, Leung HF (2012) Learning to achieve socially optimal solutions in general-sum games. In: PRICAI 2012: trends in artificial intelligence. Springer, Berlin/Heidelberg, pp 88–99
2. Hao JY, Leung HF (2014) Introducing decision entrustment mechanism into repeated bilateral agent interactions to achieve social optimality. In: Auton Agents Multi-Agent Syst 29(4):658–682
3. Hao JY, Leung HF (2012) Achieving social optimality with influencer agents. In: Proceedings of Complex'12, Santa Fe
4. Shoham Y, Powers R, Grenager T (2007) If multi-agent learning is the answer, what is the question? Artif Intell 171:365–377
5. Watkins CJCH, Dayan PD (1992) Q-learning. Mach Learn 8:279–292
6. Bowling MH, Veloso MM (2003) Multiagent learning using a variable learning rate. Artif Intell 136:215–250
7. Fudenberg D, Levine DK (1998) The theory of learning in games. MIT, Cambridge
8. Crandall JW, Goodrich MA (2005) Learning to teach and follow in repeated games. In: AAAI workshop on multiagent learning, Pittsburgh
9. Powers R, Shoham Y (2005) Learning against opponents with bounded memory. In: Proceedings of IJCAI'05, Edinburgh, pp 817–822
10. Littman ML, Stone P (2001) Leading best-response strategies in repeated games. In: IJCAI workshop on economic agents, models, and mechanisms, Seattle
11. Claus C, Boutilier C (1998) The dynamics of reinforcement learning in cooperative multiagent systems. In: Proceedings of AAAI'98, Madison, pp 746–752
12. Osborne MJ, Rubinstein A (1994) A course in game theory. MIT, Cambridge
13. Littman M (1994) Markov games as a framework for multi-agent reinforcement learning. In: Proceedings of ICML'94, New Brunswick, pp 322–328

14. Hu J, Wellman M (1998) Multiagent reinforcement learning: theoretical framework and an algorithm. In: Proceedings of ICML'98, Madison, pp 242–250
15. Banerjee D, Sen S (2007) Reaching pareto optimality in prisoner's dilemma using conditional joint action learning. In: Proceedings of AAMAS'07, Honolulu, pp 211–218
16. Sen S, Airiau S, Mukherjee R (2003) Towards a pareto-optimal solution in general-sum games. In: Proceedings of AAMAS'03, Melbourne, pp 153–160
17. Brams SJ (1994) Theory of moves. Cambridge University Press, Cambridge
18. Airiau S, Sen S (2007) Evolutionary tournament-based comparison of learning and non-learning algorithms for iterated games. J Artif Soc Soc Simul 10(3):Article no. 7
19. Hales D, Edmonds B (2003) Evolving social rationality for mas using "tags". In: Proceedings of AAMAS'03, Melbourne, pp 497–503. ACM
20. Matlock M, Sen S (2009) Effective tag mechanisms for evolving coperation. In: Proceedings of AAMAS'09, Budapest, pp 489–496
21. Matlock M, Sen S (2007) Effective tag mechanisms for evolving coordination. In: Proceedings of AAMAS'07, Honolulu, p 251
22. Hao JY, Leung HF (2011) Learning to achieve social rationality using tag mechanism in repeated interactions. In: Proceedings of ICTAI'11, Boca Raton, pp 148–155
23. Franks H, Griffiths N, Jhumka A (2013) Manipulating convention emergence using influencer agents. Auton Agents Multi-Agent Syst 26(3):315–353
24. Steels L (1995) A self-organizing spatial vocabulary. Artif Life 2(3):319–392
25. Sen S, Airiau S (2007) Emergence of norms through social learning. In: Proceedings of IJCAI'07, Hyderabad, pp 1507–1512
26. Villatoro D, Sen S, Sabater-Mir J (2009) Topology and memory effect on convention emergence. In: Proceedings of WI-IAT'09, Milano, pp 233–240
27. Bowling M, Veloso M (2002) Multiagent learning using a variable learning rate. Artif Intell 136:215–250

# Chapter 7
# Conclusion

This book mainly investigates the general question of how a desirable goal can be achieved given that each agent may have some limitations (in terms of communication and environment information) within two major multiagent interaction environments: cooperative multiagent environment and competitive multiagent environment. In cooperative multiagent environment, we focused on the solution concepts of fairness and social optimality and described a number of efficient learning strategies for agents to coordinate on fair or socially optimal outcomes for different multiagent interaction problems, which are summarized as follows:

**Fairness in cooperative multiagent environments** We introduced an adaptive strategy for achieving fairness in repeated games with conflicting interests, by incorporating the descriptive fairness model [1] in a computational way. Both theoretical and experimental results showed that behaviors of the agents are in agreement with the underlying intuition of the fairness model; therefore this strategy can be applicable to situations when the agents have to interact with humans. Besides, compared with related work, simulation results showed that agents using the adaptive strategy are able to achieve more optimal fairness results requiring limited periodical communication only, i.e., higher utilitarian social welfare under the same length of unfairness. We also considered approaching the goal of fairness from game-theoretic perspective. We first introduced the concepts of fairness strategy and fairness strategy equilibrium in the context of infinitely repeated game from the game-theoretic perspective. We showed that it provides a general and flexible way of guiding and coordinating the players to achieve both fair and highest payoffs in different types of games with conflict interest. We also described a descriptive fairness model within the game-theoretic framework which incorporates two important aspects of fairness, reciprocity and inequity aversion, in which the game-theoretic solution concept of fairness equilibrium is defined using the concept of psychological Nash equilibrium. Some general results about which outcomes can be fairness equilibria in games of different scales were presented.



J. Hao, H.-f. Leung, *Interactions in Multiagent Systems: Fairness, Social Optimality and Individual Rationality*, DOI 10.1007/978-3-662-49470-7_7

**Social optimality in cooperative multiagent environments** We first investigated the multiagent coordination problem by proposing two types of learners (IALs and JALs) in cooperative multiagent environments under the social learning framework, which is complementary to the large body of previous work in the framework of repeated interactions among fixed agents. We then extended this learning framework to be applicable for the case of general-sum games, and the performance of this social learning framework was extensively evaluated under the testbed of two-player general-sum games and better performance can be achieved compared with previous work. Finally, we considered a specific cooperative multiagent environment—the multiagent resource allocation problem through negotiation. We proposed a three-stage negotiation protocol APSOPA for multiple rational agents to achieve the socially optimal allocation through negotiation. It is theoretically proven that the final allocation is always socially optimal as long as the agents are cooperative-individually rational. We also experimentally showed the efficiency of the APSOPA protocol in terms of the reduction of both communication and computational cost.

In competitive multiagent environment, we focused on the goals of individual rationality and social optimality, and a number of strategies and mechanisms were described for achieving those goals under different multiagent interaction frameworks accordingly, which are summarized as follows:

**Individual rationality in competitive multiagent environments** We considered a specific competitive multiagent environment—bilateral negotiation—and we proposed an adaptive negotiation *ABiNeS* strategy for automated agents to negotiate to maximize its individual benefits. We introduced the concept of nonexploitation point $\lambda$ to adaptively adjust the *ABiNeS* agent's concession degree to its negotiating opponent and proposed a reinforcement learning-based approach to determine the optimal proposal for the negotiating partner to maximize the possibility that the offer will be accepted by the opponent. The performance of the *ABiNeS* strategy was evaluated using two different measures: *efficiency* (average payoff) within a single negotiation tournament and *robustness* which is determined by the size of the basin of attraction of those strategy profiles that our strategy belongs to under different negotiation tournament settings. The *ABiNeS* strategy was shown to be very efficient against the state-of-the-art strategies from ANAC 2012 and can obtain the highest average payoff over a large number of negotiation domains. Detailed analysis of the *ABiNeS* strategy in terms of the influences of its two major decision components on the negotiation efficiency was also provided, which gives us valuable insights of why it can win the championship in ANAC 2012. Last but not the least, we introduced how to apply model checking techniques to perform EGT analysis to determine the robustness of the strategies, and the *ABiNeS* strategy was found to be very robust in both bilateral negotiations and negotiation tournaments among eight players following the setting of ANAC competition.

**Social optimality in competitive multiagent environments** We first proposed a learning strategy TaFSO consisting of both *oracle* and *follower* strategies' characteristics to achieve socially optimal outcomes in competitive two-agent repeated interaction scenarios. We considered an interesting variation of sequential play by introducing an additional action $F$ for each agent. The introduction of action $F$ serves as an additional signal to facilitate the coordinate between agents, and the adoption of this signal is voluntary and determined by the agents themselves independently. Simulation results showed that a TaFSO agent can effectively influence a number of rational opponents toward socially optimal outcomes, and better performance can be achieved compared with previous work under both the case of against a class of rational learners and self-play. We also considered the problem of achieving socially optimal solutions within the social learning framework, and we proposed inserting influencer agents into the system to manipulate the behaviors of individually rational agents toward coordination on socially optimal outcomes. We showed that a small percentage of influencer agents can successfully incentivize individually rational agents to cooperate and thus achieve socially optimal outcomes.

# Reference

1. Fehr E, Schmidt KM (1999) A theory of fairness, competition and cooperation. Q J Econ 114:817–868

# Appendix A
# The 57 Structurally Distinct Games

The 57 structurally distinct games mentioned in Sect. 6.2.3 are listed as follows.

| 1's payoff, 2's payoff | | Player 2's action | |
|---|---|---|---|
| | | C | D |
| Player 1's action | C | 3, 4 | 4, 2 |
| | D | 2, 3 | 1, 1 |

A.1.1 Game 1

| 1's payoff, 2's payoff | | Player 2's action | |
|---|---|---|---|
| | | C | D |
| Player 1's action | C | 3, 4 | 4, 2 |
| | D | 1, 3 | 2, 1 |

A.1.2 Game 2

| 1's payoff, 2's payoff | | Player 2's action | |
|---|---|---|---|
| | | C | D |
| Player 1's action | C | 3, 4 | 4, 1 |
| | D | 2, 3 | 1, 2 |

A.1.3 Game 3

| 1's payoff, 2's payoff | | Player 2's action | |
|---|---|---|---|
| | | C | D |
| Player 1's action | C | 3, 4 | 4, 1 |
| | D | 1, 3 | 2, 2 |

A.1.4 Game 4

| 1's payoff, 2's payoff | | Player 2's action | |
|---|---|---|---|
| | | C | D |
| Player 1's action | C | 2, 4 | 4, 2 |
| | D | 1, 3 | 3, 1 |

A.1.5 Game 5

| 1's payoff, 2's payoff | | Player 2's action | |
|---|---|---|---|
| | | C | D |
| Player 1's action | C | 2, 4 | 4, 1 |
| | D | 1, 3 | 3, 2 |

A.1.6 Game 6

| 1's payoff, 2's payoff | | Player 2's action | |
|---|---|---|---|
| | | C | D |
| Player 1's action | C | 3, 3 | 4, 2 |
| | D | 2, 4 | 1, 1 |

A.1.7 Game 7

| 1's payoff, 2's payoff | | Player 2's action | |
|---|---|---|---|
| | | C | D |
| Player 1's action | C | 3, 3 | 4, 2 |
| | D | 1, 4 | 2, 1 |

A.1.8 Game 8

| 1's payoff, 2's payoff | | Player 2's action | |
|---|---|---|---|
| | | C | D |
| Player 1's action | C | 3, 3 | 4, 1 |
| | D | 1, 4 | 2, 2 |

A.1.9 Game 9

| 1's payoff, 2's payoff | | Player 2's action | |
|---|---|---|---|
| | | C | D |
| Player 1's action | C | 2, 3 | 4, 2 |
| | D | 1, 4 | 3, 1 |

A.1.10 Game 10

| 1's payoff, 2's payoff | | Player 2's action | |
|---|---|---|---|
| | | C | D |
| Player 1's action | C | 2, 3 | 4, 1 |
| | D | 1, 4 | 3, 2 |

A.1.11 Game 11

| 1's payoff, 2's payoff | | Player 2's action | |
|---|---|---|---|
| | | C | D |
| Player 1's action | C | 3, 4 | 4, 1 |
| | D | 2, 2 | 1, 3 |

A.1.12 Game 12



J. Hao, H.-f. Leung, *Interactions in Multiagent Systems: Fairness, Social Optimality and Individual Rationality*, DOI 10.1007/978-3-662-49470-7

| 1's payoff, 2's payoff | | Player 2's action | |
| --- | --- | --- | --- |
| | | C | D |
| Player 1's action | C | 3, 4 | 4, 1 |
| | D | 1, 2 | 2, 3 |

A.1.13 Game 13

| 1's payoff, 2's payoff | | Player 2's action | |
| --- | --- | --- | --- |
| | | C | D |
| Player 1's action | C | 3, 4 | 2, 2 |
| | D | 1, 3 | 4, 1 |

A.1.14 Game 14

| 1's payoff, 2's payoff | | Player 2's action | |
| --- | --- | --- | --- |
| | | C | D |
| Player 1's action | C | 3, 4 | 2, 1 |
| | D | 1, 3 | 4, 2 |

A.1.15 Game 15

| 1's payoff, 2's payoff | | Player 2's action | |
| --- | --- | --- | --- |
| | | C | D |
| Player 1's action | C | 3, 4 | 1, 2 |
| | D | 2, 3 | 4, 1 |

A.1.16 Game 16

| 1's payoff, 2's payoff | | Player 2's action | |
| --- | --- | --- | --- |
| | | C | D |
| Player 1's action | C | 3, 4 | 1, 1 |
| | D | 2, 3 | 4, 2 |

A.1.17 Game 17

| 1's payoff, 2's payoff | | Player 2's action | |
| --- | --- | --- | --- |
| | | C | D |
| Player 1's action | C | 2, 4 | 3, 2 |
| | D | 1, 3 | 4, 1 |

A.1.18 Game 18

| 1's payoff, 2's payoff | | Player 2's action | |
| --- | --- | --- | --- |
| | | C | D |
| Player 1's action | C | 2, 4 | 3, 1 |
| | D | 1, 3 | 4, 2 |

A.1.19 Game 19

| 1's payoff, 2's payoff | | Player 2's action | |
| --- | --- | --- | --- |
| | | C | D |
| Player 1's action | C | 3, 4 | 2, 3 |
| | D | 1, 2 | 4, 1 |

A.1.20 Game 20

| 1's payoff, 2's payoff | | Player 2's action | |
| --- | --- | --- | --- |
| | | C | D |
| Player 1's action | C | 3, 4 | 1, 3 |
| | D | 2, 2 | 4, 1 |

A.1.21 Game 21

| 1's payoff, 2's payoff | | Player 2's action | |
| --- | --- | --- | --- |
| | | C | D |
| Player 1's action | C | 2, 4 | 3, 3 |
| | D | 1, 2 | 4, 1 |

A.1.22 Game 22

| 1's payoff, 2's payoff | | Player 2's action | |
| --- | --- | --- | --- |
| | | C | D |
| Player 1's action | C | 3, 3 | 4, 1 |
| | D | 2, 2 | 1, 4 |

A.1.23 Game 23

| 1's payoff, 2's payoff | | Player 2's action | |
| --- | --- | --- | --- |
| | | C | D |
| Player 1's action | C | 3, 3 | 4, 1 |
| | D | 1, 2 | 2, 4 |

A.1.24 Game 24

| 1's payoff, 2's payoff | | Player 2's action | |
| --- | --- | --- | --- |
| | | C | D |
| Player 1's action | C | 3, 2 | 4, 1 |
| | D | 2, 3 | 1, 4 |

A.1.25 Game 25

| 1's payoff, 2's payoff | | Player 2's action | |
| --- | --- | --- | --- |
| | | C | D |
| Player 1's action | C | 3, 2 | 4, 1 |
| | D | 1, 3 | 2, 4 |

A.1.26 Game 26

| 1's payoff, 2's payoff | | Player 2's action | |
| --- | --- | --- | --- |
| | | C | D |
| Player 1's action | C | 2, 3 | 4, 1 |
| | D | 1, 2 | 3, 4 |

A.1.27 Game 27

| 1's payoff, 2's payoff | | Player 2's action | |
| --- | --- | --- | --- |
| | | C | D |
| Player 1's action | C | 2, 2 | 4, 1 |
| | D | 1, 3 | 3, 4 |

A.1.28 Game 28

| 1's payoff, 2's payoff | | Player 2's action | |
| --- | --- | --- | --- |
| | | C | D |
| Player 1's action | C | 3, 2 | 2, 1 |
| | D | 4, 3 | 1, 4 |

A.1.29 Game 29

| 1's payoff, 2's payoff | | Player 2's action | |
| --- | --- | --- | --- |
| | | C | D |
| Player 1's action | C | 2, 2 | 4, 1 |
| | D | 3, 3 | 1, 4 |

A.1.30 Game 30

| 1's payoff, 2's payoff | | Player 2's action | |
|---|---|---|---|
| | | C | D |
| Player 1's action | C | 2, 2 | 3, 1 |
| | D | 4, 3 | 1, 4 |

A.1.31 Game 31

| 1's payoff, 2's payoff | | Player 2's action | |
|---|---|---|---|
| | | C | D |
| Player 1's action | C | 2, 2 | 4, 1 |
| | D | 1, 4 | 3, 3 |

A.1.32 Game 32

| 1's payoff, 2's payoff | | Player 2's action | |
|---|---|---|---|
| | | C | D |
| Player 1's action | C | 3, 4 | 4, 3 |
| | D | 1, 2 | 2, 1 |

A.1.33 Game 33

| 1's payoff, 2's payoff | | Player 2's action | |
|---|---|---|---|
| | | C | D |
| Player 1's action | C | 3, 4 | 4, 3 |
| | D | 2, 2 | 1, 1 |

A.1.34 Game 34

| 1's payoff, 2's payoff | | Player 2's action | |
|---|---|---|---|
| | | C | D |
| Player 1's action | C | 2, 4 | 4, 3 |
| | D | 1, 2 | 3, 1 |

A.1.35 Game 35

| 1's payoff, 2's payoff | | Player 2's action | |
|---|---|---|---|
| | | C | D |
| Player 1's action | C | 3, 4 | 4, 3 |
| | D | 2, 1 | 1, 2 |

A.1.36 Game 36

| 1's payoff, 2's payoff | | Player 2's action | |
|---|---|---|---|
| | | C | D |
| Player 1's action | C | 3, 4 | 4, 3 |
| | D | 1, 2 | 2, 1 |

A.1.37 Game 37

| 1's payoff, 2's payoff | | Player 2's action | |
|---|---|---|---|
| | | C | D |
| Player 1's action | C | 3, 4 | 4, 2 |
| | D | 2, 1 | 1, 3 |

A.1.38 Game 38

| 1's payoff, 2's payoff | | Player 2's action | |
|---|---|---|---|
| | | C | D |
| Player 1's action | C | 3, 4 | 4, 2 |
| | D | 1, 1 | 2, 3 |

A.1.39 Game 39

| 1's payoff, 2's payoff | | Player 2's action | |
|---|---|---|---|
| | | C | D |
| Player 1's action | C | 3, 3 | 4, 2 |
| | D | 2, 1 | 1, 4 |

A.1.40 Game 40

| 1's payoff, 2's payoff | | Player 2's action | |
|---|---|---|---|
| | | C | D |
| Player 1's action | C | 3, 3 | 4, 2 |
| | D | 1, 1 | 2, 4 |

A.1.41 Game 41

| 1's payoff, 2's payoff | | Player 2's action | |
|---|---|---|---|
| | | C | D |
| Player 1's action | C | 2, 4 | 4, 1 |
| | D | 3, 2 | 1, 3 |

A.1.42 Game 42

| 1's payoff, 2's payoff | | Player 2's action | |
|---|---|---|---|
| | | C | D |
| Player 1's action | C | 2, 4 | 3, 1 |
| | D | 4, 2 | 1, 3 |

A.1.43 Game 43

| 1's payoff, 2's payoff | | Player 2's action | |
|---|---|---|---|
| | | C | D |
| Player 1's action | C | 2, 3 | 4, 1 |
| | D | 3, 2 | 1, 4 |

A.1.44 Game 44

| 1's payoff, 2's payoff | | Player 2's action | |
|---|---|---|---|
| | | C | D |
| Player 1's action | C | 2, 3 | 3, 1 |
| | D | 4, 2 | 1, 4 |

A.1.45 Game 45

| 1's payoff, 2's payoff | | Player 2's action | |
|---|---|---|---|
| | | C | D |
| Player 1's action | C | 3, 4 | 2, 1 |
| | D | 4, 2 | 1, 3 |

A.1.46 Game 46

| 1's payoff, 2's payoff | | Player 2's action | |
|---|---|---|---|
| | | C | D |
| Player 1's action | C | 3, 3 | 2, 1 |
| | D | 4, 2 | 1, 4 |

A.1.47 Game 47

| 1's payoff, 2's payoff | | Player 2's action | |
|---|---|---|---|
| | | C | D |
| Player 1's action | C | 2, 3 | 4, 2 |
| | D | 1, 1 | 3, 4 |

A.1.48 Game 48

| 1's payoff, 2's payoff | | Player 2's action | |
|---|---|---|---|
| | | C | D |
| Player 1's action | C | 2, 4 | 4, 1 |
| | D | 1, 2 | 3, 3 |

A.1.49 Game 49

| 1's payoff, 2's payoff | | Player 2's action | |
|---|---|---|---|
| | | C | D |
| Player 1's action | C | 2, 4 | 4, 3 |
| | D | 1, 1 | 3, 2 |

A.1.50 Game 50

| 1's payoff, 2's payoff | | Player 2's action | |
|---|---|---|---|
| | | C | D |
| Player 1's action | C | 3, 4 | 2, 1 |
| | D | 1, 2 | 4, 3 |

A.1.51 Game 51

| 1's payoff, 2's payoff | | Player 2's action | |
|---|---|---|---|
| | | C | D |
| Player 1's action | C | 2, 4 | 3, 1 |
| | D | 1, 2 | 4, 3 |

A.1.52 Game 52

| 1's payoff, 2's payoff | | Player 2's action | |
|---|---|---|---|
| | | C | D |
| Player 1's action | C | 2, 3 | 3, 4 |
| | D | 4, 2 | 1, 1 |

A.1.53 Game 53

| 1's payoff, 2's payoff | | Player 2's action | |
|---|---|---|---|
| | | C | D |
| Player 1's action | C | 2, 2 | 3, 4 |
| | D | 4, 3 | 1, 1 |

A.1.54 Game 54

| 1's payoff, 2's payoff | | Player 2's action | |
|---|---|---|---|
| | | C | D |
| Player 1's action | C | 2, 2 | 4, 3 |
| | D | 3, 4 | 1, 1 |

A.1.55 Game 55

| 1's payoff, 2's payoff | | Player 2's action | |
|---|---|---|---|
| | | C | D |
| Player 1's action | C | 2, 4 | 4, 2 |
| | D | 1, 1 | 3, 3 |

A.1.56 Game 56

| 1's payoff, 2's payoff | | Player 2's action | |
|---|---|---|---|
| | | C | D |
| Player 1's action | C | 3, 3 | 2, 4 |
| | D | 4, 2 | 1, 1 |

A.1.57 Game 57

Printed in the United States
By Bookmasters